Transgenic Maize

Series Editor
John M. Walker
School of Life Sciences
University of Hertfordshire
Hatfield, Hertfordshire, AL10 9AB, UK

For other titles published in this series, go to
www.springer.com/series/7651

METHODS IN MOLECULAR BIOLOGY™

Transgenic Maize

Methods and Protocols

Edited by

M. Paul Scott

USDA-ARS, Corn Insects and Crop Genetics Research Unit, Ames, IA, USA

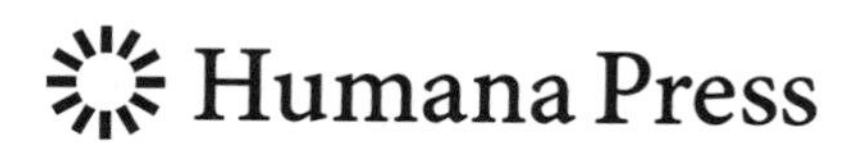

Editor
M. Paul Scott
USDA-ARS
Corn Insects and Crop Genetics Research Unit
1407 Agronomy Hall
Ames, IA 50011

ISBN: 978-1-934115-49-7 e-ISBN: 978-1-59745-494-0
ISSN: 1064-3745 e-ISSN: 1940-6029
DOI: 10.1007/978-1-59745-494-0

Library of Congress Control Number: 2008943239

Printed on acid-free paper

springer.com

Preface

Maize is unique among plants because it is one of the most important crops in the world and it is one of the foremost model species for genetic studies. Maize grain is important as a staple food crop in many parts of the world and is also a major component of livestock and poultry diets. Maize is also grown as a forage crop for ruminant animals. In addition, maize is emerging as an important contributor to fuel markets. Maize grain is used to make ethanol and maize stover has been proposed to be a source of biomass for lignocellulosic biofuel production. As a genetic model, experiments in maize led to the rediscovery of Mendel's law's of inheritance *(1)*, quantitative genetics theory *(2)*, characterization of effect of X-rays on chromosomes *(3)*, and the discovery of transposable elements *(4)*. Increasing global population and economic development ensure continued need for production of maize and other crops.

Several features of maize have led to its success as a model species and crop plant. First, it has tremendous genetic diversity. It is in part this diversity that has allowed plant breeders to develop maize varieties that are adapted to a broad range of environmental conditions. Maize is monoecious, meaning each plant has separate male and female flowers. This aids the making of controlled pollinations, facilitating breeding and genetic experiments. Finally, maize possesses C4 physiology, enabling it to fix carbon very efficiently.

Transgenic technology allows the transfer of genetic material across species boundaries or the use of genetic material that does not exist in nature. This ability is valuable in genetic research. It allows genetic complementation studies and manipulation of maize systems on a molecular level. It has also been used as a crop improvement tool. Transgenic maize that is resistant to herbicides reduces crop management effort and expense, while transgenic maize producing insecticidal proteins increases yield and quality of maize grain. While not in widespread commercial production yet, the scientific literature contains examples of successful manipulation of maize nutritional quality and functional properties. It is possible that transgenic technology will play a role in improving draught tolerance and nitrogen use efficiency, two traits that limit crop production in many areas.

One of the challenges of working with transgenic maize is that it requires expertise in a large number of disciplines. Transgenic maize projects are often carried out as cooperative efforts between research teams or use fee-for-service laboratories to accomplish parts of the project objectives. This book is roughly divided by the disciplines required to take a transgenic maize experiment from idea to evaluation of transgenic plants. Thus, design of transgenes requires a thorough understanding of gene expression and cell biology and is addressed in the first chapter. In the next section, several methods of maize transformation are described. Following this, examples of the use of transgenic maize in research are given. Expression of reporter genes, down-regulation of gene activity, and gene isolation are broadly applicable research objectives. A chapter covers each of these areas, focusing on the fluorescent proteins, RNAi, and plasmid rescue technologies that have emerged from several other technologies as being broadly effective. The next section is devoted to methods of analysis of transgenic maize plants at levels ranging from DNA to transgene function. The final section addresses issues encountered at the level of whole plants and describes experiments that would typically be carried out in a field or greenhouse setting. In the United States of America, the Animal and Plant Health Inspection Service (APHIS) regulates the release of experimental transgenic varieties to the environment. Appendices are included with examples of the information required by APHIS. While it is unlikely that any given reader will use all of the procedures in this book, it is hoped that the reader will benefit from the procedures in this book that they use and that other procedures are helpful for understanding parts of the experiments being performed by cooperators or fee-for-service labs.

M. Paul Scott
Ames, IA

References

1. Correns, C. G. (1900) Mendel's Regel über das Verhalten der Nachkommenschaft der Rassenbastarde. *Berichte Deut. Bot. Ges.* 18, 158–168.
2. Emerson, R. A. and East, E. M. (1913) The inheritance of quantitative characters in maize. *Bull. Agric. Exp. Stn. Nebr.* 2, 1–120.
3. McClintock, B. (1950) The origin and behavior of mutable loci in maize. *Proc. Natl Acad. Sci. USA* 36, 344–355.
4. Stadler L.J. (1928) Genetic effects of x-rays in maize. *Proc. Natl Acad. Sci. USA*, 14, 69–75.

Contents

Part V: Breeding with Transgenes

Contributors

NICOLE L. ARNOLD • *Discovery Research, Cell Biology, Dow Agrosciences LLC, Indianapolis, IN, USA*
EARL BICAR • *Monsanto, Kihei, HI, USA*
DAVID DUNCAN • *Monsanto, Chesterfield, MO, USA*
BRONWYN FRAME • *Center for Plant Transformation, Iowa State University, Ames, IA, USA*
DAVE JACKSON • *Plant Genetics, Cold Spring Harbor Laboratory, Cold Spring Harbor, NY, USA*
DONG LIU • *Department of Biological Sciences and Biotechnology, Tsinghua University, Beijing, China*
ANDING LUO • *Plant Genetics, Cold Spring Harbor Laboratory, Cold Spring Harbor, NY, USA*
KAREN MCGINNIS • *Plant Sciences, University of Arizona, Tucson, AZ, USA*
AMITABH MOHANTY • *DuPont Knowledge Centre, Hyderabad, India*
ADRIENNE N. MORAN LAUTER • *Corn Insects and Crop Genetics Research Unit, USDA-ARS, Ames, IA, USA*
GUO-LING (GILLIAN) NAN • *Biological Sciences, Stanford University, Stanford, CA, USA*
JOAN PETERSON • *BASF, Ames, IA, USA*
JOSEPH PETOLINO • *Discovery Research, Cell Biology, Dow Agrosciences LLC, Indianapolis, IN, USA*
M. PAUL SCOTT • *Corn Insects and Crop Genetics Research Unit, USDA-ARS, Ames, IA, USA*
COLIN T. SHEPHERD • *AimsBio, Inc., Ames, IA, USA*
VLADIMIR SIDEROV • *Monsanto, Chesterfield, MO, USA*
ANNE SYLVESTER • *Department of Molecular Biology, University of Wyoming, Laramie, WY, USA*
KARLA VOGEL • *Monsanto, Ankeny, IA, USA*
VIRGINIA WALBOT • *Biological Sciences, Stanford University, Stanford, CA, USA*
KAN WANG • *Center for Plant Transformation, Iowa State University, Ames, IA, USA*
YAN YANG • *Plant Genetics, Cold Spring Harbor Laboratory, Cold Spring Harbor, NY, USA*

Part I

Introduction

Chapter 1

Design of Gene Constructs for Transgenic Maize

Dong Liu

Summary

The first step of any maize transformation project is to select gene expression elements that will make up an effective construct. When designing a gene construct, one must have a full understanding of the different expression elements that are currently available and of the strategies that have been successfully used to overcome obstacles in past. In this chapter, we discuss several major classes of expression elements that have been used for maize transformation, including promoters, introns, and untranslated regions. We also discuss several strategies for further improving transgene expression levels, such as optimization of codon usage, removal of deleterious sequences, addition of signal sequences for subcellular protein targeting, and use of elements to reduce position effects. We hope that this chapter can serve as a general guideline to help researchers, especially beginners in the field, to design a gene construct that will have the maximum potential for gene expression.

Keywords: Transgenic maize, Gene construct, Promoters, Introns, UTRs, Codon bias

1. Introduction

To successfully express a foreign gene in transgenic maize, it is critical to properly select the expression elements that are used to control both the "gene of interest" and the selectable marker gene. The expression level of an introduced gene in transgenic maize can be affected by many factors, including promoter activity, stability of mRNA, protein translation efficiency, protein stability in a particular cellular compartment, the chromosomal position of the integrated gene, and degree of gene silencing. This chapter will discuss how these factors should be considered and how to properly select expression-controlling elements when designing a gene construct for maize transformation.

M. Paul Scott (ed.), *Methods in Molecular Biology: Transgenic Maize, vol. 526*

DOI: 10.1007/978-1-59745-494-0_1

Because data reported in the literature are not always consistent, care must be taken when incorporating some features into the gene construct. The selection criteria for the expression elements discussed here may be used only for a general maize transformation project. More sophisticated techniques, such as generation of marker-free transgenic plants, and the use of homologous recombination systems for site-specific gene integration, are beyond the scope of this chapter and can be found in some review articles if one is interested *(1, 2)*.

2. An Overview of Plant Gene Expression

In plant genomes, a functional gene unit generally consists of a DNA segment which encodes a protein, and regulatory DNA elements which control the expression of the protein coding region. **Figure 1** illustrates the steps that are involved in the process of plant gene expression. The first step of gene expression is the transcription of DNA to pre-mRNA, which is initiated by the regulatory element called the promoter. It is usually located at the 5′ end of the gene, upstream of the transcription start site. The promoter contains various sequence elements that function in the recruitment of protein factors that facilitate transcription of the protein coding region of the gene. The most basic elements are the TATA box and CAAT box, the names of which are based on their nucleotide sequences. Transcription termination is determined by a terminator sequence, which is located downstream of the translation termination site.

Once the pre-mRNA is transcribed from its DNA template, a few steps remain before the mRNA is mature. Splicing removes intron sequences that do not encode protein. A 7-methylguanosine nucleotide cap is added to the 5′ end. Near the 3′ end, a specific sequence, typically AAUAAA, is recognized by cellular machinery, generating a mature 3′ end by cleavage about 20–30 nucleotides away from the motif and the addition of a poly (A) tail. Since the terminator contains the sequence motif for polyadenylation and the whole 3′ untranslated region, or 3′ UTR, the terms terminator, 3′ polyadenylation sequence, and 3′ UTR have been used interchangeably in the context of vector construction.

Thus, the mature mRNA molecule consists of a protein coding region and UTR at both the 5′ and 3′ ends, with a 5′ cap and 3′ poly A tail. Finally, the mature mRNA will be translated into a protein. The resulting protein will be further modified, folded into the proper configuration, and targeted into a specific cellular compartment.

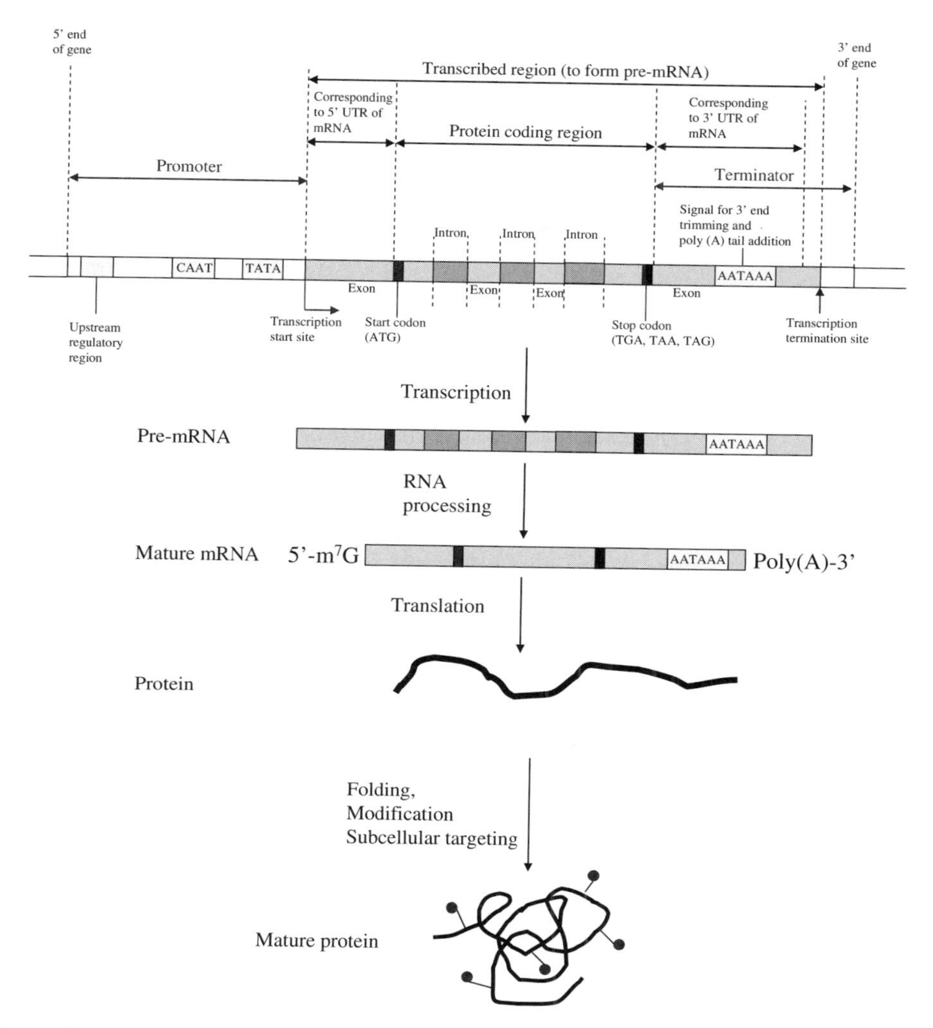

Fig. 1. An overview of the process of plant gene expression.

3. Promoters and Introns

The promoter is the most critical element for controlling gene expression. Promoters can be divided into three general categories: constitutive, tissue-specific, and inducible. A constitutive promoter drives expression of a gene in all plant tissues and throughout all developmental stages. A tissue-specific promoter expresses the gene in only a specific type of tissue. Tissue-specific promoters may or may not cause expression in all developmental stages. An inducible promoter only initiates gene expression under certain external conditions, such as light, temperature, nutrient levels, or in response to application of a particular chemical.

In the early days, the 35S promoter from Cauliflower Mosaic Virus (*CaMV 35S*) was most commonly used to drive high-level gene expression in plant tissue *(3)*. However, further study showed that the activity of *CaMV 35S* promoter in monocot plants is much lower than the activity in dicot plants. To improve its activity in monocots, several modifications have been made, including the duplication of the enhancer region of the promoter *(4)*. Another is the insertion of an intron found in various promoters of cereal plants. The intron is usually inserted between the 35S promoter and the initial codon of the transgene coding region.

Introns that have been tested include those from the *Adh1*, *Sh1*, *Ubi1*, and *Act1* genes from maize, and the *Chs* gene from petunia *(5)*. Among them, the intron from the maize *Ubi1* promoter gives the highest level of enhancement. The intron from the *Chs* gene gives a surprising 100-fold enhancement. These introns are all quite long (about 1 kb) and located within the 5′ UTRs in their native genes. The mechanism of enhancement produced by these introns in improving gene expression is still not well understood. It has been generally thought that inclusion of an intron can improve efficiency of mRNA processing and increase the steady-state levels of mRNA *(6)*. Even though the activity of unmodified *CaMV 35S* promoter is lower in monocots than in dicots, it has still been routinely used in maize transformation, when combined with an intron or some UTR sequences, for expressing screenable or selectable marker genes *(7)*.

So far, the constitutive promoter most commonly used to drive a "gene of interest" in transgenic maize is the promoter of maize *Ubi1* combined with the first intron located in its 5′ UTR region *(8)*. As with the examples mentioned earlier, the inclusion of this intron greatly enhances the level of transgene expression. ProdiGene, a plant biotechnology company in Texas, has used the *Ubi1* promoter to express avidin and beta-glucuronidase (GUS) genes in transgenic maize, both of which are used as diagnostic agents in molecular biology. In highly expressive transgenic lines, avidin constitutes over 2% of the aqueous protein extracted from dry maize seeds while the recombinant GUS protein can accumulate to 0.7% of total soluble proteins *(9, 10)*. These two products from transgenic maize are now commercially available from Sigma-Aldrich. It has been estimated that the cost of producing avidin in transgenic maize is about only 10% of the cost of extraction from egg white.

ProdiGene also used the *Ubi1* promoter to drive four transgenes that are the components of a secretory antibody – heavy chain, light chain, joining chain, and a secretory component *(11)*. In this construct, each gene is expressed from the same *Ubi* promoter. In these transgenic maize seeds, they could detect the assembled antibodies with expression levels of up to 0.3% of total soluble protein in T1 seeds. Based on ProdiGene's success with other products, the selection of high-performance lines and backcrossing should allow this yield to be increased as much as 70-fold over six generations.

Besides *CaMV 35S* and *Ubi1* promoters, the promoter of the maize histone *H2B* gene has also proven capable of driving constitutively high-level expression of a transgene, based on the recovered rate of transformants and histochemical analysis of reporter gene expression *(12)*. However, there are no data reported on the quantitative side-by-side comparison of *H2B*, *Ubi1*, and other promoters in maize plants.

In transgenic rice, rice *Act1* is the most commonly used constitutive promoter to drive high-level expression of a transgene *(13, 14)*. The structure and function of this promoter has been well characterized in rice plants. However, for unknown reasons, this promoter has not been extensively exploited in transgenic maize.

Constitutive promoters are very useful in directing expression of selectable or screenable maker genes, because screening requires the expression of these genes being initiated as soon as possible after integration into plant genome. This ensures that selection for the transformants can be effective and can reduce the false positive rate. There are other uses for a strong constitutive promoter besides marker genes. Herbicide-resistance genes allow breeders to easily follow the presence of the transgene in a plant population and create the valuable trait of crop resistance to weed-control chemicals. Constitutive expression of insecticidal proteins is useful if the target pest is one that feeds on all or most plant parts.

In some cases, specific promoters have advantages over strong constitutive promoters. For example, confining the production of a pharmaceutical recombinant protein to a certain tissue type could prevent unwanted dispersal of the protein in pollen grains. Tissue-specific expression could also aid in easier harvesting and extraction of the protein of interest. Localized expression via a tissue-specific promoter can prevent unnecessary depletion of plant nutrients when the protein of interest is not needed in all parts of the plant. This is applicable when using insecticidal proteins to deter pests that attack only certain parts of the plant.

There has been some research into using tissue-specific promoters for pest control in transgenic maize. Johnson et al. *(15)* used a putative silk-specific promoter to drive the expression of a maize Myb transcription factor in transgenic maize, enhancing resistance to corn earworm, which feeds on corn silks. Corn leaf aphid is a sap-sucking pest, which feeds on phloem tissues. Using a phloem-specific promoter of rice sucrose synthase gene, Wang et al. *(16)* expressed snowdrop lectin in transgenic maize. The expression level of snowdrop lectin ranged from 0.13 to 0.28%

of total soluble protein. Three lines with expression levels above 0.22% showed significant resistance to aphids.

In transgenic maize, the most commonly used tissue-specific promoters are for the expression of recombinant protein in seeds. Advantages of expressing protein in seeds include confinement of the transgene product, easy storage prior to protein extraction, and high-level expression. Various seed-specific promoters have been used in maize, many derived from seed storage protein genes. These include promoters from maize 27-kDa zein (*zmZ27*), maize *waxy* (starch synthase) genes, rice *glutelin-1*, and the small subunit of ADP-glucose pyrophosphorylase gene, all of which are maize endosperm-specific *(17, 18)*. The embryo-specific *globulin-1* promoter has also been used to express recombinant proteins in maize *(19)*.

Recently, Yu et al. *(20)* used maize seed storage protein promoter *P19z* to drive the expression of the lysine-rich pollen-specific protein Sb401 from potato in maize. Their results showed that the expression level of Sb401 was correlated with increased levels of lysine and total protein content in maize seeds. In their R1 maize seeds, compared with the nontransgenic maize control, the lysine content increased by 16.1–54.8% and the total protein content increased by 11.6–39.0%. Further analysis showed that the levels of lysine and total protein remained high for six continuous generations, indicating that the elevated lysine and total protein levels were heritable. However, care must be taken because although these promoters are often described as seed-specific promoters, a low level of activity can also be detected in other plant tissues. For example, maize *waxy* promoter that is endosperm-preferred also shows low activity in pollen *(17)*.

An alternative way to produce recombinant proteins in transgenic maize is to use maize cell culture. Cell culture confers the advantages of confinement of the transgene and its product. The short life cycle in cell culture allows for faster harvesting of the protein of interest. In addition to constitutive promoters, inducible promoters may be considered for this purpose. In rice cell culture, it has been demonstrated that the rice *alpha-amylase* promoter can drive high-level gene expression during sugar starvation *(21)*. Recently, the same authors used this promoter combined with the signal peptide sequence of the *amylase* gene to express human serum albumin (HSA) in rice cell culture *(22)*. Upon sugar starvation, the highest yield of HSA that is secreted into the culture medium can reach 74 mg/L. This expression system may be adopted for the production of recombinant pharmaceutical proteins in maize cell culture. To date, no research has been conducted on this subject.

In some cases, expression of a transgene can have detrimental effects on the host cells. Inducible promoters have the advantage of tight control over the timing of transgene expression. These promoters allow expression, only after exposure to the proper chemical. As with tissue-specific promoters, inducible promoters avoid or reduce unnecessary consumption of nutrients and energy in transgenic plants or cultured cells. Several chemically regulated plant promoters have been reported in the literature *(23)*. None of them has been adopted for field use, mainly due to government regulation of chemicals released into the environment. Some chemicals, such as the antibiotics tetracycline *(24)* and the steroid dexamethasone, are not suitable to be used in large quantities in the field, though dexamethasone has been widely used as an inducer in laboratory conditions *(2)*.

4. Untranslated Regions

Regulation of mRNA stability and protein translation efficiency is just as important as promoter choice when considering levels of gene expression. Many studies have shown that the 5′ and 3′ UTRs on the transcript can play an important role in gene expression levels *(25)*. The presumed role of the 5′ UTR is to influence the efficiency with which the bound ribosomal subunits migrate and recognize the translation start codon. A useful 5′ UTR for transgene expression must be able to efficiently assemble ribosomes and present the start codon in an appropriate configuration. Generally, an AT-rich 5′ UTR allows the ribosomal complex to easily scan to the start codon to initiate translation.

When the 5′ UTR from CaMV 35S was used with its promoter in maize protoplasts, it enhanced reporter gene expression about 40-fold *(26)*. The 5′ UTRs from the maize genes *glutelin* *(27)* and *PEP-carboxylase* *(28)* have also been tested in maize protoplasts, resulting in 12-fold and 3.7-fold increased expression, respectively. Effects of a given 5′ UTR on the transgene expression can be very different in monocots and dicots, and not every 5′ UTR has an enhancing effect. For example, the 5′ UTRs from genomic RNA of tobacco mosaic virus and alfalfa mosaic virus coat protein have been particularly effective in enhancing transgene expression in tobacco *(29, 30)*, but they are much less effective in maize cells *(31)*. In contrast, the 5′ UTR from a tobacco etch virus has been effectively used to drive a selectable marker gene in maize plants when combined with the *CaMV 35S* promoter *(7, 32)*.

There is evidence that the 3′ UTR determines stability of mRNA as well as facilitates polyadenylation of the mRNA. A 3′ UTR sequence with a strong stem-loop structure can effectively block degradation of mRNA by RNase that starts from the 3′ end of the transcript. Ingelbrecht et al. *(33)* examined the effects of 3′

UTRs from several different genes, finding that their effects on transgene expression vary widely. The different 3′ ends used for testing were from the octopine synthase gene on the Ti plasmid of *Agrobacterium tumefaciens*, the 2S seed protein gene from Arabidopsis, extensin gene from carrot, the gene of the small subunit of *rbcS* from Arabidopsis, and chalcone synthase gene from Antirrhinum. In stable tobacco transformants, there was about a 60-fold difference between the best-expressing construct (with the 3′ UTR from small subunit of *rbcS*) and the lowest-expressing construct (with the 3′ UTR from a chalcone synthase), with the other 3′ UTRs producing different expression levels between these two extremes.

A popular choice of the 3′ UTR for both monocot and dicot plants, including maize, is the one from the nopaline synthase (*nos*) gene of *Agrobacterium* Ti plasmid *(34)*. This small unit (about 260 bp) is probably the most widely used 3′ UTR today in transgenic plants. Other commonly used 3′ UTRs include those from *CaMV 35S* transcripts *(7)*, the potato *pinII* gene *(35)*, and the soybean vegetative storage protein gene *(36)*. To date, no comparative studies on the effects of different 3′ UTRs on the transgene expression in transgenic maize plants have been reported.

Inclusion of the UTR sequences from plant genes in constructs for expression of bacterial genes in transgenic plants is particularly useful. In our previous work, we attached the 5′ and 3′ UTR sequences from a tobacco osmotin gene to both ends of the full-length *Toxin A* gene from bacterium *Photorhabdus luminescens*, and several of its derivatives *(37)*. The *Toxin A* gene is 7.5-kb long, encoding a 283-kD insecticidal protein. Achieving expression of Toxin A in plant cells is very difficult. With the attachment of the UTR sequences from the osmotin gene, expression levels of Toxin A protein in Arabidopsis plants were increased 10-fold and the recovery rate of insect-resistant plants increased 12-fold. We chose the UTR sequences from tobacco osmotin because both osmotin mRNA and protein are very stable in plant cells. Computer modeling of its 3′ UTR showed many stem-loop structures, which may effectively block degradation by RNAse (Liu et al. unpublished). The attachment of plant UTR sequences makes this prokaryotic gene behave more like a "plant gene," so corresponding mRNA can be easily recognized by plant translation machinery. Homologs of osmotin exist in almost every monocot plant, including maize *(38)*. It will be interesting to test if the UTR sequences from the maize osmotin homolog gene will have the same enhancing effect on gene expression in transgenic maize.

The sequence that immediately surrounds the translation start codon is another important factor that influences the translation efficiency. Kozak *(39)* compared vertebrate mRNA sequences to the results of site-directed mutagenesis, and demonstrated the existence of a preferred nucleotide context surrounding the initial codon. Kozak has defined an optimal AUG context for vertebrates as GCC(A/G)CCAUGG (the start codon is underlined), named the Kozak consensus sequence. The most highly conserved position of the consensus is position −3 (the A of the AUG codon being +1) where 97% of vertebrate mRNAs have a purine. The −3 position appears to mediate the efficiency of translation initiation in vertebrate systems. Such a consensus sequence has also been found among plant genes: UAACAAUGGCU *(40, 41)*. In position +4, the preference for a guanine is significantly high in plants (85%). The preference for G in position +4 (85%) and C at +5 (77%) in plants resembles the preference for A in position −3 in animals. Further compilation of sequences surrounding the AUG from 85 maize genes yields a consensus of (C/G)AUGGCG *(42)*. This consensus sequence should be taken into consideration when making gene constructs for use in maize.

5. Codon Bias

During early attempts to express a foreign gene in transgenic plants, it was frequently found that only a very low amount of the foreign protein accumulated. This is particularly true when introducing a prokaryotic gene into plant cells. The example that best illustrates this point is of the expression of insecticidal crystal proteins of *Bacillus thuringiensis (Bt)*. When Vaeck et al. *(43)* first introduced *cryIA(b)* gene into plant cells, they found that the expressed foreign protein was only 0.0002–0.02% of total soluble protein. Similar results were reported by other researchers. Subsequent analysis of these transgenic plants showed that mRNA species were shorter than the expected full-length transcript, suggesting inefficient posttranscriptional processing or rapid turnover of full-length transcripts *(44)*.

Close examination of many bacterial native endotoxin genes has found that they tend to have a very low G+C content, around 37%. In contrast, plant genes in general tend to have a higher G+C content, with maize showing a strong preference for G+C-rich coding regions *(45, 46)*. The A+T-rich sequence in bacterial genes may provide many putative motifs, which will function in plant cells as splice sites, polyadenylation signal sequences *(40, 44, 47)*, and RNA destabilizing elements, such as ATTTA. The short motif of ATTTA was first found in animal cells, causing mRNA instability when present in the the 3′ UTR of the transcript *(48)*. This phenomenon has been confirmed in plant cells *(49)*. Since all plant introns contain A/T-rich sequences *(70)*, the presence of AT-rich elements, such as four or more consecutive A or T bases, may facilitate aberrant pre-mRNA splicing. With the presence of these deleterious sequence motifs, early constructs for Bt in plant cells produced highly unstable mRNA, resulting in very little translation of intact protein.

Codon bias is another major obstacle preventing expression of bacterial genes in transgenic plants. Different species may exhibit a preference or bias for one of many synonymous codons for a given amino acid. For the amino acid arginine in alfalfa, the codon CGU is 50 times more likely to occur in the genome than the rarest codon, CGG. In maize, AGG and CGC are the preferred codons for arginine, but the codon CGG is also frequently used. **Table 1** lists the codon frequencies in the maize genome.

Many codons commonly found in maize are rare in bacteria, and vice versa. Codon frequency for two insecticidal protein

Table 1
Codon usage of 2,280 coding sequences of maize genes

Amino acid	Codon	Freq	Amino acid	Codon	Freq	Amino acid	Codon	Freq	Amino acid	Codon	Freq
Ala	GCU	21.1	Gln	CAA	13.3	Leu	TTA	5.7	Ser	TCG	10.5
Ala	GCC	31.2	Gln	CAG	23.5	Leu	CTG	25.8	Ser	TCA	11.2
Ala	GCA	16.7	Glu	GAG	40.9	Leu	CTA	7.3	Thr	ACC	16.5
Ala	GCG	23.1	Glu	GAA	20.0	Lys	AAG	39.6	Thr	ACT	10.8
Arg	AGG	14.8	Gly	GGT	14.3	Lys	AAA	15.0	Thr	ACA	10.5
Arg	CGC	14.3	Gly	GGC	30.2	Met	ATG	24.1	Thr	ACG	10.8
Arg	AGA	8.8	Gly	GGA	13.3	Phe	TTC	25.1	Trp	TGG	12.9
Arg	CGT	6.1	Gly	GGG	15.3	Phe	TTT	12.6	Tyr	TAC	19.3
Arg	CGG	9.4	His	CAC	14.8	Pro	CCA	13.9	Tyr	TAT	9.4
Arg	CGA	4.3	His	CAT	10.1	Pro	CCT	12.6	Val	GTC	21.1
Asn	AAC	22.2	Ile	ATC	23.0	Pro	CCC	13.5	Val	GTG	25.6
Asn	AAT	13.5	Ile	ATT	14.0	Pro	CCG	15.4	Val	GTT	15.8
Asp	GAC	32.2	Ile	ATA	8.4	Ser	TCC	16.2	Val	GTA	6.3
Asp	GAT	23.0	Leu	CTC	25.5	Ser	TCT	12.1	Ter	TGA	1.1
Cys	TGC	12.1	Leu	TTG	13.2	Ser	AGC	16.1	Ter	TAA	0.5
Cys	TGT	5.6	Leu	CTT	15.9	Ser	AGT	7.8	Ter	TAG	0.7

This table is adopted from the following website with some modifications: http://www.kazusa.or.jp/codon/cgi-bin/showcodon.cgi?species = Zea + mays + [gbpln] *Freq*: occurred frequency per thousand codons genes (*tcdA* and *tcbA*) from the bacterium *P. luminescens* can be found in an issued patent (Patent No. US 6,590,142 B1). The codons GCG for alanine, GAA for glutamate, and CTG for leucine were not used in these two bacterial genes, but their frequencies in maize are about 23, 20, and 26% in each amino acid group, respectively.

The low expression levels correlated with codon bias are probably due to the small pool size of the specific tRNAs that are encoded by the rare codons. Presence of these rare codons could cause ribosomes to stall during translation of the transcripts. Stalled ribosomes may cause mRNA destabilization by leaving the transcripts exposed to components of the RNA degradation machinery.

Besides the average codon usage for amino acids, attention should also be paid to some other unique features in coding sequences of plant genes. Intercodon nucleotide doublets CG and AT are rarely seen in plant genes *(46)*. In contrast, CT and TG doublets frequently occur. The potential RNA polymerase II termination sequence CAN7-9AGTNNA, though not extensively studied, is believed to be important for transcription termination *(50)*. Furthermore, although maize genes usually have codons with G or C in the third position, a majority of codons in endotoxin genes have A or T in the third position. Thus, the presence of sequences deleterious to mRNA stability in plant cells, coupled with differences in codon bias, can make bacterial genes extremely difficult to express in plant cells.

Using a completely synthesized coding sequence is an effective way to solve the problem of transgene expression. By reengineering the whole coding sequence, one can not only change the codon preference and elevate the overall G+C content that is close to the level of the targeting plant species, but also can remove all the deleterious sequence motifs that may cause mRNA instability. **Table 2** lists some sequence motifs that should be avoided during the gene rebuilding process. Through gene rebuilding, people have made significant improvement in expression of endotoxin genes in cotton *(51)*, tomato *(52)*, and potato *(53)*.

The first synthetic gene that was successfully expressed in transgenic maize is the *cryIA(b)* gene from Bt var. Kurstaki HD-1 *(54)*. The coding region for the first 648 amino acids of the 1,155 amino acid CryIA(b) was synthesized using the most preferred codons from maize for each amino acid *(55)*. The synthetic gene had 65% homology with the native gene and a G+C content of 65%, compared with 37% of the native gene. When compared with the native gene, the synthetic gene produced significantly higher level of CryIA(b) protein in both tobacco and maize. In maize plants, the protein produced from native gene was not detectable, while the synthetic gene produced up to 4 ug/mg soluble protein in certain plants.

Table 2
Sequence motifs that should be avoided in synthetic bacterial gene expression in plants *(44, 47, 48)*

RNA destabilizing element	Polyadenylation signal sequence	Splice site	Intercodon doublet	Nucleotide in 3rd position	AT-rich sequence	RNAP II
ATTTA	AATAAA[a]	GT	CG	A and T	(A) 4 or more	CAN7-9-AGTNNA
	AATTAT[b]	AG	AT		(T) 4 or more	
	AGTATA[b]					
	AATATT[b]					
	AATGAA[c]					
	AAAATA[c]					
	AAGCAT[c]					
	AATAAT[c]					
	AATAAG[c]					
	ATACAT[c]					
	ATTAAT[c]					

RNAP II: RNA polymerase II termination sequence
[a]The most commonly used polyadenylation signal sequence in both monocots and dicots
[b]Three sequence motifs found in bacterial endotoxin genes that also act as polyadenylation signals in plants
[c]Potential polyadenylation signal sequences found in some monocot genes. There are other polyadenylation signal sequences of dicot plants not listed in this table

With the great success of using synthetic genes to express bacterial proteins in transgenic plants, the gene rebuilding process has become a routine practice in transgenic plant programs.

6. Subcellular Targeting

Protein stability is another key factor in regulating the expression level of a heterologous protein in transgenic plants. The cellular compartment in which a protein accumulates can strongly influence the interrelated process of folding, assembly, and posttranslational modification. Also, when trying to modify a biosynthetic pathway, it is important that the transgene product accumulates in the same compartment as the metabolites that will serve as substrate. For these reasons, subcellular targeting must be considered in order to obtain a desired trait in transgenic plants.

In plants, it has been clear that entry of a protein into the endoplasmic reticulum (ER) is mediated by a signal sequence of about 20 amino acids, generally located at its N-terminus. If no further signaling determinants are present in the protein, it will be loaded into the secretory pathway. The default destination for this pathway is the intercellular matrix or cell culture medium *(56)*. In previously mentioned examples, ProdiGene used the maize *ubi1* promoter combined with a signal peptide from barley alpha-amylase to express avidin and *Trametes versicolor* laccase 1 isoform in maize seeds. They found that expression levels of each protein could reach up to 2% and 1%, respectively, and expressed recombinant protein accumulated in intercellular space *(9, 19)*. High accumulation of the proteins in the intercellular space may be due to the lack of proteinase in these locations.

Another benign environment for protein accumulation in plant cells is the lumen of the ER. The mechanism by which proteins are retained in the ER has been well characterized and shown to be evolutionarily conserved among plants and animals. A tetrapeptide motif, usually KDEL or HDEL, present at the C-terminus of the protein, is necessary and sufficient to cause ER retention, or at least recycling back to the ER *(57)*. There are various reports that proteins targeted to the ER by addition of a KDEL or HDEL sequence at its C-terminus can accumulate to relatively high levels in tobacco plants *(58–60)*.

Retention of protein in the ER lumen has other advantages as well. The activity of many proteins is dependent on the posttranslational modification, such as the formation of disulfide bonds and glycosylation. These modification processes all occur in the lumen of the ER. Retention of foreign proteins in the ER lumen can also prevent the proteins from being transported into the protease-rich environment of the vacuole.

So far, this ER targeting strategy has not been vigorously exploited in transgenic maize. In an attempt to improve nutrition of maize with porcine alpha-lactalbumin, Yang et al. *(61)* made three gene constructs with a synthetic protein coding sequence containing maize preferred codons. All three constructs used the *ubi1* promoter and *nos* terminator for controlling gene expression. The first construct has no additional amino acid sequences attached to the porcine alpha-lactalbumin gene. The second and third constructs have signal peptide sequences added at the 5′ end of the gene. One of the two has a KDEL sequence attached to its C-terminus. The transgenic plants with the first construct did not produce detectable levels of recombinant protein, indicating the importance of a signal peptide for protein accumulation. Both the second and third constructs produced recombinant protein in callus and kernels, but the data did not allow them to distinguish between the two constructs regarding the level of protein produced or the frequency of expression.

7. Position Effect

It is well known that expression levels of transgenes in primary transformants can vary substantially. The degree of variability differs among gene constructs and plant species, but even expression from the same construct in the same plant host can have considerable variation. Random placement of a transgene, known as the position effect, is the main explanation for high intertransformant variability. Transgene expression can be influenced by the properties of the surrounding genomic DNA: the presence of flanking promoters, enhancing or silencing DNA elements, the copy number of integrated transgene, the chromatin structure or methylation status, etc. Today, there is no transformation technology that can effectively control where the transgene is integrated into the chromosomes of the targeted plants. Thus, it is still a costly and time-consuming process to sort a large number of individual transformants to find one or several events that may have a desired trait with stable inheritability for several generations.

Reduced intertransformant variability has been reported with the use of special elements such as matrix attachment regions (MAR) or scaffold-associated regions (SARs) *(62–64)*. It has also been reported that certain MARs provide expression levels in transgenic rice that are higher and more predictable *(65, 66)*. However, the use of different MARs or SARs, in combination with various promoters, can have different effects on gene expression in different systems *(67)*. In transgenic maize, Sidorenko et al. *(68)* tested the effects of two MARs, one from the *p1- rr* gene and another from the *Adh1* gene, on the variability of transgene expression when combined with three different promoters. Their results showed that these two MARs have no effects on the reduction of variability of transgene expression; however, the Adh1-MAR did modify the spatial pattern of expression driven by one synthetic promoter. In another report, Torney et al. *(69)* tested two MARs from the *Adh1* gene, revealing new transgene expression patterns in roots and systemically induced variegation, but no effects on reduction of transgene expression variability among the individual transformants. So far, no effective methods have been developed in transgenic maize to eliminate the position effect that causes the variability of transgene expression. The only way to obtain high-performance transgenic lines is to generate large numbers of transformants, and then select individuals with the desired trait through sorting.

In summary, the success of a maize transformation project is largely dependent on a construct design that incorporates the proper expression elements and strategies. Figure 2 is a diagram of the decision-making tree. We hope that this chapter will help

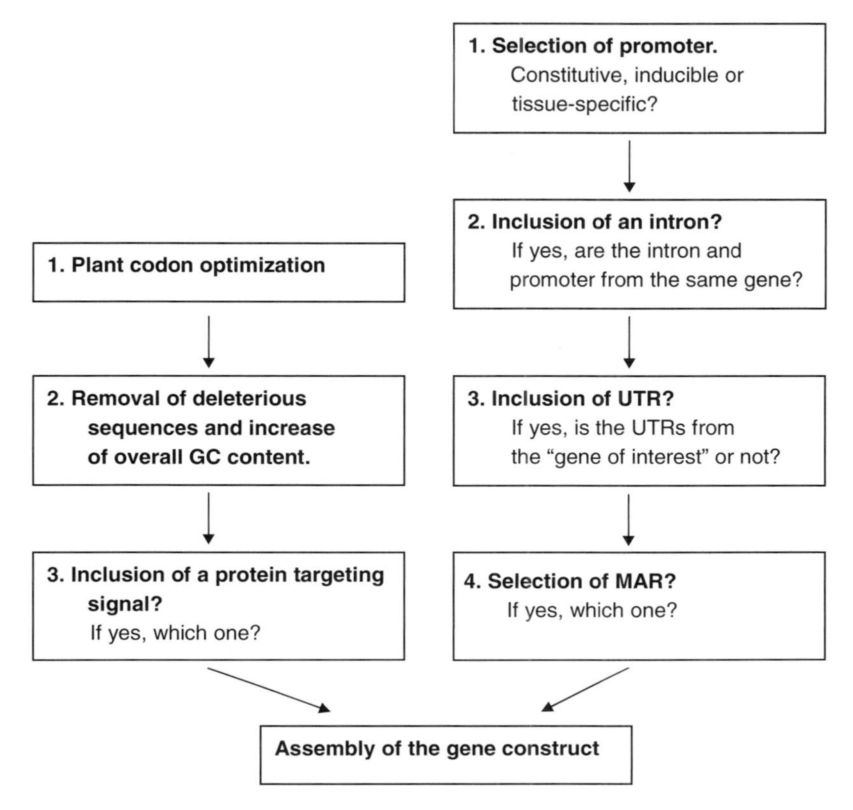

Fig. 2. A flow chart of the decision-making process for the design of a gene construct for maize transformation.

researchers design gene constructs that will have the maximum potential for expression of transgenes in maize.

Acknowledgments

The author sincerely wishes to thank Anastasia Bodnar for her critical reading and help on the manuscript preparation.

Note

The research on insect-resistant Arabidopsis plants with Toxin A gene (**ref.** *37*) was conducted at Dow AgroScience LLC, 9330 Zionsville, IN 46268, U.S.A.

References

1. Darbani, B., Eimanifar, A., Stewart, C.N. Jr., and Camargo, W.N. (2007) *Methods to produce marker-free transgenic plants. Biotechnology* 2(1), 83–90.
2. Zuo, J., Niu, Q.W., Ikeda, Y., andChua N.H. (2002) *Marker-free transformation: increasing transformation frequency by the use of regeneration-promoting genes. Curr. Opin. Biotechnol.* 13(2), 173–180.
3. Odell, J.T., Nagy, F., and Chua, N.H. (1985) *Identification of DNA sequence required for activity of the cauliflower mosaic virus 35S promoter. Nature* 313, 810–812.
4. Kay, R., Chan, A., Daly, M., and McPherson, A. (1987) *Duplication of CaMV 35S promoter sequence creates a strong enhancer for plant genes. Science* 236, 1299–1302.
5. Vain, P., Finer, K.R., Engler, D.E., Pratt, R., and Finer J. (1996) *Intron-mediated enhancement of gene expression in maize (Zea mays L.) and bluegrass (Poa pratensis L.). Plant Cell Rep.* 15, 489–494.
6. Lueharsen, K.R., and Walbot, V. (1991) *Intron enhancement of gene expression and splicing efficiency of introns in maize cells. Mol. Gen. Genet.* 225, 81–93.
7. Frame, B.R., Shou, H., Chikwamba, R.K., Zhang, Z., Xiang, C., Fonger,T.M., Pegg, S.E., Li, B., Nettleton, D.S., Pei, D., and-Wang K. (2002) *Agrobacterium tumefaciens-mediated transformation of maize embryos using a standard binary vector system. Plant Physiol.* 129(1), 13–22.
8. Christensen, A.H., and Quail, R.H. (1996) *Ubiquiting promoter-based vectors for high-level expression of selectable and/or screenable marker genes in monocotyledons plants. Transgenic Res.* 5, 231–218.
9. Hood, E., Witcher, D., Maddock, S., Meyer, T., Baszczynski, C., et al. (1997) *Commercial production of avidin from transgenic maize: characterization of transformant, production, processing, extraction and purification. Mol. Breed.* 3, 291–306.
10. Witcher, D.R., Hood, E.E., Peterson, D., Bailey, M., Bond, D., et al. (1998) Commercial production of β-glucuronidase (GUS): a model system for the production of proteins in plants. Mol. *Breed.* 4, 301–312.
11. Hood, E.E., Woodard, S.L., and Horn, M. (2002) *Monoclonal antibody manufacturing in plants: myths and realities. Curr. Opin. Biotechnol.* 13, 630–635.
12. Rasco-Gaunt, L.S., Li, C., Hagemann, K., Riley, A., et al. (2003) *Characterization of the expression of a novel constitutive maize promoter in transgenic wheat and maize. Plant Cell Rep.* 21, 569–576.
13. McElroy, D., Zhang, W., Cao, J., and Ray, W. (1990) *Isolation of an efficient actin promoter for use in rice transformation. Plant Cell* 2(2), 163–171.
14. Wang, Y., Zhang, W., Cao, J., and Ray, W. (1992) *Characterization of cis-acting elements regulating transcription from the promoter of a constitutively active rice actin gene. Mol. Cell Biol.* 12(8), 399–406.
15. Johnson, E., Berhow, M., and Dowd, P. (2007) *Expression of a maize Myb transcription factor driven by a putative silk-specific promoter significantly enhances resistance to Helicoverpa zea in transgenic maize. J. Agric. Food Chem.* 55, 2998–3003.
16. Wang, Z., Zhang, K., Sun, X., Tang, K., and Zhang, J. (2005) *Enhancement of resistance to aphids by introducing the snowdrop lectin gene into maize plants. J. Biosci.* 30, 627–638.
17. Russell, D.A., and Fromm, M.E. (1997) *Tissue-specific in transgenic maize of four endosperm promoters from maize and rice. Transgenic Res.* 6, 157–168.
18. Zheng, Z.W., Kawagoe, Y., Xiao, S.H., Li, Z., and Okita, T. (1993) *5 distal and proximal cis-acting regulator elements are required for developmental control of a rice seed storage protein glutelin gene. Plant J.* 4, 357–366.
19. Hood, E.E., Bailey, M.R., Beifuss, K., Magallane-Lundback, M., and Callaway, E. (2003) *Criteria for high-level expression of a fungal laccase gene in transgenic maize. Plant Biotechnol. J.* 1, 129–140.
20. Yu, J., Peng, P., Zhang, X., Zhao, Q., Zhu, D., Sun, X., Liu, J., and Ao, G. (2005) Food Nutr. *Bull.* 26, 427–431.
21. Chan, M., Chao, Y., and Yu, S. (1994) *Novel gene expression system for plant cells based on induction of alpha-amylase promoter by carbohydrate starvation. J. Biochem. Chem.* 269, 17635–17641.
22. Huang, L., Liu, Y., Lu, C., Hsieh, S., and Yu, S. (2005) *Production of human serum albumin by sugar starvation induced promoter in rice cell culture. Transgenic Res.* 14, 569–581.
23. Gatz, C. (1997) *Chemical control of gene expression. Annu. Rev. Plant Physiol. Plant Mol. Biol.* 48, 89–108.
24. Weinmann, P., Gossen, M., Hillen., W., Bujard, H., and Gatz, C. (1994) *A chimeric transactivator allows tetracycline-responsive gene expression in whole plants. Plant J.* 5, 559–569.

25. Gallie, D., and Bailey-Serre, J. (1997) *Eyes off transcription! The wonderful world of post-transcriptional regulation. Plant Cell* 9, 667–673.
26. Rothstein, S.J., Lahners, K.N., Lostein, R.J, Carozzi, N.B., and Jayne, S.M. (1987) *Rice DA: promoter cassettes, antibiotic-resistance genes, and vectors for plant transformation. Gene* 53, 153–161.
27. Boronat, A., Martinez, M.C., Reina, M., Puigdomenech, P., and Palau, J. (1986) *Isolation and sequencing of a 28 kd glutenin-2 gene from maize: common elements in the 5′ flanking regions among zein and glutenin genes. Plant Sci.* 47, 95–102.
28. Hudspeth, L.W., and Grula, J.W. (1989) *Structure and expression of the maize gene encoding the phosphoenolpyruvate carboxylase isozyme involved in C4 photosynthesis. Plant Mol. Biol.* 12, 579–589.
29. Dowson, D., Ashurst, M.J., Mathias, J.L., Watts, S.F., Wilson, T.M., and Dixon, R.A. (1993) *Plant viral leaders influence expression of a reporter gene in tobacco. Plant Mol. Biol.* 23, 97–109.
30. Jobling, S.A., and Gehrke, L. (1987) *Enhanced translation of chimeric messenger RNAs containing a plant viral untranslated leader sequence. Nature* 325, 622–625.
31. Gallie, D.R., and Young, T.E. (1994) *The regulation of expression in transformed maize aleurone and endosperm protoplast. Plant Physiol.* 106, 929–939.
32. Carrington, J.C., and Freed, D.D. (1990) *Cap-independent of enhancement of translation by a plant poty virus 5′ non-translated region. J. Virol.* 64, 1590–1597.
33. Ingelbrecht, L.W., Herman, L.M.F., Dekeyser, R.A., Van Montagu, M.C., and Depicker, A.G. (1989) *Different 3′ end regions strongly influence the level of gene expression in plant cells. Plant Cell* 1, 671–680.
34. Depicker, A., Stachel, S., Dhaese, P., Zambryski, P., and Goodman, H.M. (1982) *Nopaline synthase, transcript mapping and DNA sequence. J. Mol. Appl. Genet.* 1, 561–573.
35. An, G., Mitra, A., Hong, K.C., Costa, M.A., An, K., and Thornburg, R.W., et al. (1989) *Functional analysis of the 3′ control region of the potato wound-inducible proteinase inhibitor II gene. Plant Cell* 1, 115–122.
36. Mason, H.S., DeWald, D., and Mullet, J.E. (1993) *Identification of a methyl jasmonate-responsive domain in the soybean vspB promoter. Plant Cell* 15, 241–251.
37. Liu, D., Burton, S., Glancy, T., Li, Z., Hampton, R., Meade T., and Merlo, D. (2003) *Insect resistance conferred by 283-kD Photorhabdus luminescens protein TcdA in Arabidopsis thaliana.* Nat. *Biotechnol.* 21, 1222–1228.
38. Vigers, A.J., Roberts, W.K., and Selitrennikoff, C.P. (1991) *A new family of plant antifungal proteins. Mol. Plant. Microbe Interact.* 4, 315–323.
39. Kozak, M. (1986) *Point mutations define a sequence flanking the AUG initiator codon that modulates translation by eukaryotic ribosomes. Cell* 44, 283–292.
40. Joshi, C.P. (1987) *Putative polyadenylation signals in nuclear genes of higher plants: a compilation and analysis. Nucleic Acids Res.* 15, 9627–9640.
41. Lutcke, H.A., Chow, K.C., Mickel, F.S., Moss, K.A., Kern, H.F., and Scheele, G.A. (1987) *Selection of AUG initiation codon differs in plants and animals. EMBO J.* 6, 43–48
42. Lueharsen, K.R., and Walbot, V. (1994) *The impact of AUG start codon context on maize gene expression in vivo. Plant Cell Rep.* 13, 454–458.
43. Vaeck, M., Reynaerts, A., Hofte, H., Jansens, S., De Beuckeleer, M., Dean, C., et al. (1987) *Transgenic plants protected from insect attack. Nature* 328, 33–37.
44. Diehn, S.H., Chiu, W-L., De Rocher, E.J., and Green, P.J. (1998) *Prematuration polyadenylation at multiple sites within a Bacillus thuringiensis toxin gene-coding region. Plant Physiol.* 117, 1433–1443.
45. Cambell, C.H., and Gowri, G. (1990) *Codon usage in higher plants, green algae, and cyanobacteria. Plant Physiol.* 92, 1–11.
46. Murray, E.E., Lotzer, J., and Eberle, M. (1989) *Codon usage in plants. Nucleic Acids Res.* 17, 477–498.
47. Dean, C., Tamaki, S., Dunsmuir, P., Favreau, M., Katayama, C., et al. (1986) *mRNA transcripts of several plant genes are polyadenylated at multiple sites in vivo. Nucleic Acids Res.* 14, 2229–2240.
48. Shaw, G., and Kamen, R. (1986) *A conserved AU sequence from the 3′ untranslated region of GM-CSF mRNA mediates selective mRNA degradation. Cell* 46, 659–667.
49. Ohme-Takagi, M., Taylor, C. B., Newman, T. C., andGreen, P. (1993) *The effect of sequences with high AU content on mRNA stability in tobacco. Proc. Natl Acad. Sci. USA.* 90, 11811–11815.
50. Vankan, P., and Filipowicz, W. (1988) *Structure of U2 snRNA genes of Arabidopsis thaliana and their expression in electroporated protoplast. EMBO J.* 7, 791–799.

51. Perlak, F.J., Deaton, R.W., Amstrong, T.A., Fuchs, R.L., Sims, S.R., Greenplate, J.T., and Fischhoff, D.A. (1990) *Insect resistant cotton plants. Bio/Technology* 8, 939–943.
52. Perlak, F.J., Fuchs, R.L., Dean, D.A., McPherson, S.L., and Fischhoff, D.A. (1991) *Modification of the coding sequence enhances plant expression of insect control protein genes. Proc. Natl Acad. Sci. USA.* 88, 3324–3328.
53. Adang, M.J., Brody, M.S., Cardineau, G., Eagan, N., Roush, R.T., and Shewmaker, C.K. et-al. (1993) *The reconstruction and expression of a Bacillus thuringiensis cryIIIA gene in protoplast and potato plants. Plant Mol. Biol.* 21, 1131–1145.
54. Geiser, M., Schweitzer, S., and Grimm, C. (1986) *The hypervariable region in the genes coding for entomopathogenic crystal proteins of Bacillus thuringiensis: nucleotide sequence of the kurhd1 gene of subsp. Kurstaki HD1. Gene* 48, 109–118.
55. Koziel, M.G., Beland, G.L., Bowman, C., Carrozzi, N.B., Crenshaw, R., et al. (1993) *Field performance of elite transgenic maize plants expressing an insecticidal protein derived from Bacillus thuringiensis. Bio/Technology.* 11, 194–200.
56. Denecke, J., Botterman, J., and Deblaere, R. (1990) *Protein secretion in plant cells can occur via a default pathway. Plant Cell.* 2, 531–550.
57. Munro, S., and Pelham, H.R.B. (1987) *A C-terminal signal prevents secretion of luminal ER proteins. Cell* 48, 899–907.
58. Napier, R.M., Fowke, L.C., Hawes, C., Lewis, M., and Pelham, H.R.B. (1992) *Immunological evidence that plants use HDEL and KDEL for targeting proteins to the endoplasmic reticulum. J. Cell Sci.* 102, 261–271.
59. Schouten, A., Roosien, J., van Engelen, F.A., de Jong, G.A.M., Borst-Vrenssen, A.W.M. et-al. (1996) *The C-terminal KDEL sequence increases the expression of a single-chain antibody designed to be targeted to both the cytosol and the secretory pathway in transgenic tobacco. Plant Mol. Biol.* 30, 781–793.
60. Wandelt, C.I., Khan, M.R.I., Craig, S., Schroeder, H.H., Spencer, D., and Higgins, T.J.V. (1992) *Vicilin with carboxy-terminal KDEL is retained in the endoplasmic reticulum and accumulates to high levels in the leaves of transgenic plants. Plant J.* 2, 181–192.
61. Yang, S.H., Moran, D.L., Jia, H.W., Bicar, E.H., Lee, M., and Scott, M.P. (2002) *Expression of a synthetic porcine alpha-lactabumin gene in the kernels of transgenic maize. Transgenic Res.* 11, 11–20.
62. Breyne, P., Van Montagu, M., Depicker, A., and Gheysen, G. (1992) *Characterization of a plant scaffold attachment region in a DNA fragment that normalizes transgene expression in tobacco. Plant Cell* 4, 463–471.
63. Mlynarova, L., Loonen, A., Heldens, J., Jansen, R.C., Keizer, P., Stiekema, W.J., and Nap, J.P. (1994) *Reduced position effect in mature transgenic plants conferred by the chicken lysozyme matrix-associated region. Plant Cell* 6(3), 417–426.
64. Mlynarova, L., Jansen, R.C., Conner, A.J., Stiekema, W.J., and Nap, J.P. (1995) *The MAR-mediated reduction in position effect can be uncoupled from copy number-dependent expression in transgenic plants. Plant Cell* 7(5), 599–609.
65. Oh, S., Jeong, J., Kim, E., Yin, N., Jang, I., et al. (2005) *Matrix attachment region from chicken lysosome locus educes variability in transgene expression and confer copy number-dependent in transgenic rice plants. Plant Cell Rep.* 24, 145–154.
66. Xue, H., Yang, Y.T., Wu, C.A., Yang, G.D., Zhang, M.M., and Zheng, C.C. (2005) *TM2, a novel strong matrix attachment region isolated from tobacco, increases transgene expression in transgenic rice calli and plants. Theor. Appl. Genet.* 110, 620–627.
67. Mankin, S., Allen, G., Phelam, A., Spiker, S., and Tompson, W. F. (2003) *Elevation of transgene expression level by flanking matrix attachment region (MAR) is promoter dependent: a study of the interactions of six promoters with the RB7 3′ MAR. Transgenic Res.* 12, 3–12.
68. Sidorenko, L., Bruce, W., Maddock, S., Tagliani, L., Li, X., Daniels, M., and Peterson, T. (2003) *Functional analysis of two matrix attachemnt region (MAR) elements in transgenic maize plants. Transgenic Res.* 12, 137–154.
69. Torney, F., Partier, A., Saya-lesage, V., Nadaud, I., Barret, P., and Beckert, M. (2004) *Heritable transgene expression pattern imposed onto maize ubiquitin promoter by maize adh-1 matrix attachment regions: tissue and developmental specificity in maize transgenic plants. Plant Cell Rep.* 22, 931–938.
70. Goodall, G., and Filipowicz, W. (1989) *The AU-rich sequence present in the introns of plant nuclear pre-mRNA are required for splicing. Cell* 58, 473–483.

Part II

Transformation Methods

Chapter 2

Transient Expression of GFP in Immature Seed Tissues

Colin T. Shepherd

Summary

Transient expression is the nonstable expression of a transgene whereby the transgene does not integrate into the host's genomic DNA. Transient expression assays have 20 years of history in plant molecular biology research, being used to answer a variety of questions. The method described here allows the ability to test promoter activity for seed-specific expression by quantifying reporter protein production in immature seed tissues. This method is especially suited to test vector activity for stable expression, test promoter activity, and discern regions of a promoter that is necessary for transcription in seed tissues. The transient expression assay is a tool that has aided a great deal of molecular biology research.

Keywords: Transient expression, Promoter, Green fluorescent protein

1. Introduction

Transient expression is referred to as nonstable gene expression, whereby a gene is introduced into a tissue and expressed only for a short defined period of time without being stably integrated into the genome. Transient expression developed out of a need for rapid analysis of gene expression. Transient expression assays are different from stable expression of a transgene in the following ways. First, the DNA to be expressed is not integrated into the genome. Transiently expressed genes are not subject to position effects that can influence the level of transcription and the tissue specificity of genome-integrated transgenes *(1)*. Second, transient expression assays are fast, as gene expression can be measured in hours or days after DNA introduction rather than the weeks, months, or years with stable transformation studies *(2)*. Third, transient expression assays are very flexible with a

M. Paul Scott (ed.), *Methods in Molecular Biology: Transgenic Maize, vol. 526*

DOI: 10.1007/978-1-59745-494-0_2

wide variety of DNA introduction methods and a vast number of tissue types used for study *(3)*. For the study of molecular biology in plants, transient expression assays have developed into an extremely valuable tool.

Early reports of transient expression assays were focused on testing gene constructs intended for stable transformation using electroporation of DNA into plant protoplasts *(4–7)*. Transient expression allowed researchers to examine expression of heterologous genes in plants that were not transformable by *Agrobacterium tumefaciens*, such as cereals. Subsequent reports expanded upon these initial experiments, and transient expression was used for a number of research purposes such as promoter activity analysis of the maize α-zein *(8)*, enhancing transcription levels *(9)*, and dissecting promoters to identify DNA regulatory sequences *(10)*. These reports illustrate that transient expression assays can be used on a diverse range of problems, and pave the way for further research and development for better and more useful transient expression analyses.

Current transient expression assays are used in several avenues of research, including evaluation of promoter transcription activity, promoter sequence dissection and function *(11)*, reporter gene activity *(12)*, subcellular localization of proteins *(13)*, stable transformation methods of crops *(14)*, and small- and large-scale recombinant protein production *(15)*. Promoters typically consist of core regulatory domains such as TATA box and transcription initiation sites. In general, most regulatory elements that affect transcription rates are upstream of the promoter. Therefore, the region of the promoter containing the TATA box and the upstream region that contains regulatory elements are usually tested in this transient expression assay. The promoter region can be fused with a reporter gene to determine transcriptional activity of the promoter by measuring expression levels of the reporter gene in the transient expression assay. Green fluorescent protein (GFP) is used extensively as a reporter and requires no substrates for fluorescent activity, and expression levels are simply monitored by measuring the fluorescent levels in selected plant tissues *(16)*. Improved GFP is ideal for current plant and animal molecular biology research. Many of these variants have been developed using transient expression assays. Other reporter genes that have been used in transient expression assays are *β*-glucuronidase (GUS) and luciferase.

Although transient expression is used to test promoter and is also used for basic research, it has become a method to express heterologous proteins of value for recovery in plants. Expression of heterologous proteins in plants by transient expression has used two methods: viral vector-based transient amplification and *Agrobacterium* delivery of genes for transient expression *(17)*. In most cases, heterologous proteins are produced by integrating the gene into the genome; however, using transient expression, the genes do not get integrated into the genome, yet expression levels remain high. Transient expression for heterologous protein production has a few advantages such as the following: speed (3–14 days for full expression level), the gene is not integrated into the genome and therefore heritable transgenic plants are not created, and brief high-level expression can be achieved.

Determining promoter activity by measuring GFP fluorescence level is an important method to quantify transient expression results. This chapter describes a biolistic transient expression method for testing promoter activity in seed tissues such as the embryo and endosperm. This method is a quantitative measurement of promoter activity and is especially useful for testing promoters that will be used in stable transformations or testing promoter regions for activity in seed tissues.

2. Materials

1. Petri dishes, 100 mm × 15 mm.
2. Immature corn ears, 13–17 days after pollination (DAP).
3. Murashigue and Skoog salt media and Phytagar (Invitrogen, Carlsbad, CA).
4. Gene gun and supplies (gold particles and screens) (Bio-Rad, Hercules, CA).
5. Plant expression vector containing GFP gene.
6. 96-Well Spectrofluorometer (Tecan, Mannedorf/Zurich, Switzerland).
7. GFP extraction buffer: 30 mM Tris–HCl pH 8.0, 10 mM EDTA, 10 mM NaCl, and 5 mM DTT.

3. Methods

An efficient method for quantitatively testing promoters is necessary to effectively understand promoter activity and select promoters that have high activity in seed tissues. Nonquantitative methods while able to discern regions of promoters that are necessary for transcription to occur are not able to determine promoter activity levels. This protocol describes qualitative transient expression and involves quantifying the amount of GFP in each immature tissue fragment and using ANOVA statistics to characterize the variation. Negative control plasmids should be included. Negative control plasmids should consist of a DNA vector without a reporter gene.

This method is based on biolistic bombardment of seed tissues for basic research of promoter structure. Although transient expression has been used for recombinant protein expression, this method will not be suited for that end. This method does not use viral-vector technology, nor does it use *Agrobacterium* to transfer the genes, but uses DNA-coated gold particles bombarded into tissues as the method of delivery.

3.1. Transient Expression Assay

1. Prepare 13–17 DAP maize kernels from self-pollinated ears of desired genotype by hand-dissecting embryo and endosperm tissues (*see* **Note 1**). The immature ears were dehusked and placed in 5–10% bleach solution to sterilize the outer side of all tissues for 15 min and then washed with ddH_2O.
2. Set 24 immature embryos and 24 immature endosperms (hand-dissected to approximately 50 mg each) of each line in Petri dishes containing Murashigue and Skoog salt media and Phytagar. The immature tissues should be arranged on the plate in concentric circles from the center of the plate so that each tissue piece is equidistant from the piece next to it. This gives each tissue piece an equal chance of being bombarded by the gold particles. Special emphasis should be placed on determination of the field of bombardment of the gene gun. Each gene gun may have a different pattern of bombardment, and tissue pieces should be arranged so that each tissue is within that field of bombardment.
3. For each construct, biolistically bombard the tissue using 1.5-μg DNA of each plasmid. The bombardment should be repeated three times with each Petri dish so that enough cells are impacted with DNA for the assay. Incubate the bombarded tissues for 48 h at 27°C in the dark to allow GFP to accumulate. Before incubation the Petri dishes should be rapped in parafilm so that no moisture will escape during the incubation process.
4. To extract and quantify GFP, each embryo and endosperm sample must be macerated separately by mashing in 200 μL of GFP extraction buffer containing 30 mM Tris–HCl, 10 mM EDTA, 10 mM NaCl, and 5 mM DTT (*see* **Note 2**). Many techniques could be used for mashing of the tissue, such as micro pestles and 1.5-mL centrifuge tube or a 48-well tissue masher. Key to this process is the maceration of the tissue so that the majority of the cells expel their contents.
5. After centrifuging, the supernatant should be removed and placed into a black 96-well plate (*see* **Note 3**). The fluorescence level of each supernatant should be determined using a spectrofluorometer at excitation and emission wavelength that is appropriate for the reporter gene used.

4. Notes

1. When preparing immature tissues, use sterile techniques to limit contamination. Wear gloves and work in the laminar flow hood or under a flame when handling the tissues.
2. GFP extraction buffer should be chilled to 4°C before use.
3. Appropriate statistical analysis is the key to obtaining quantitative values for transient expression assays. Experiments must be designed carefully to allow this. Individual transient expression assays yield results with a high degree of variability, so many replications are required to obtain statistically significant results. We suggest 24 replications here, but this number could vary with the experimental conditions. It is important to randomize replications within a 96-well plate. If possible, design experiments so that all the samples fit on one plate to avoid plate-to-plate variation. Data can be analyzed with an ANOVA to determine the probability that the observed variation is due to chance. If the resulting F test is significant, the means can be compared using a Student's T test.

References

1. Jones J, Dunsmuir P, Bedbrook J. (1985) High level expression of introduced chimeric genes in regenerated transformed plants. *EMBO J* 4: 2411–2418.
2. Dekeyser R, Claes B, De Rycke R, Habets M, Van Montagu M, Caplan A. (1990) Transient gene expression in intact and organized rice tissues. *Plant Cell* 2: 591–602.
3. Fischer R, Vaquero-Martin C, Sack M, Drossard J, Emans N, Commandeur U. (1999) Towards molecular farming in the future: transient protein expression in plants. *Biotechnol Appl Biochem* 30: 113–116.
4. Fromm M, Taylor L, Walbot V. (1985) Expression of genes transferred into monocot and dicot plant cells by electroporation. *Proc Natl Acad Sci USA* 82: 5822–5828.
5. Fromm M, Callis J, Taylor L, Walbot V. (1987) Electroporation of DNA and RNA into plant protoplasts. *Methods Enzymol* 153: 351–366.
6. Hauptmann RM, Ozias-Akins P, Vasil V, Tabaeizadeh Z, Rogers SG, Horsch RB, Vasil IK, Fraley RT. (1987) Transient expression of electroporated DNA in monocotyledonous and dicotyledonous species. *Plant Cell Rep* 6: 265–270.
7. Junker B, Zimney J, Luhrs R, Lorz H. (1987) Transient expression of chimeric genes in dividing and non-dividing cereal protoplasts after PEG-induced DNA uptake. *Plant Cell Rep* 6: 329–332.
8. Thompson G, Boston R, Lyznik L, Hodges T, Larkins B. (1990) Analysis of promoter activity from an alpha-zein gene 5′ flanking sequence in transient expression assays. *Plant Mol Biol* 15: 755–764.
9. Maas C, Laufs J, Grant S, Korfhage C, Werr W. (1991) The combination of a novel stimulatory element in the first exon of the maize *Shrunken-1* gene with the following intron 1 enhances reporter gene expression up to 1000-fold. *Plant Mol Biol* 16: 199–207.
10. Hamilton D, Roy M, Rueda J, SindhuR, Sanford J, Mascarenhas J. (1992) Dissection of a pollen-specific promoter from maize by transient transformation assays. *Plant Mol Biol* 18: 211–218.
11. Marzabal PB, Busk PK, Ludevid MD, Torrent M. (1998) The bifactorial endosperm box of the gamma-zein gene: characterization and function of the Pb3 and GZM cis-acting elements. *Plant J* 16: 41–52.
12. Schenk P, Elliot A, Manners J. (1998) Assessment of transient gene expression in plant tissues using the green fluorescent protein as a reference. *Plant Mol Biol Rep* 16: 313–322.

13. Pollmann S, Neu D, Lehmann T, Berkowitz O, Schafer T, Weiler E. (2006) Subcellular localization and tissue specific expression of amidase 1 form *Arabidopsis thaliana*. *Planta* 224: 1241–1253.
14. Lorence A, Verpoorte R. (2004) Gene transfer and expression in plants. *Methods Mol Biol* 267: 329–350.
15. Chung S, Vaidya M, Tzfira T. (2006) Agrobacterium is not alone: gene transfer to plants by viruses and other bacteria. *Trends Plant Sci* 11: 1–4.
16. Southward C, Surette M. (2002) The dynamic microbe: green fluorescent protein brings bacteria to light. *Mol Microbiol* 45: 1191–1196.
17. Gleba Y, Klimyuk V, Marillonnet S. (2005) Magnifection – a new platform for expressing recombinant vaccines in plants. *Vaccine* 23: 2042–2048.

Chapter 3

Biolistic Gun-Mediated Maize Genetic Transformation

Kan Wang and Bronwyn Frame

Summary

Biolistic gun-mediated transformation is one of the two most effective and popular methods for introducing genes into maize. In this chapter, we describe a detailed protocol for genetic transformation of the maize genotype, Hi II. The protocol uses 0.6-µm gold particles as microcarriers and the herbicide resistance *bar* gene as a selective marker. Both immature zygotic embryos and immature embryo-derived callus cultures can be transformed using this protocol. To ensure successful reproduction of this protocol, we provide step-by-step laboratory transformation procedures as well as details on growing and caring for transgenic plants in the greenhouse.

Keywords: *Zea mays*, Maize, Genetic transformation, Biolistic gun

1. Introduction

Biolistic gun (or particle gun)-mediated DNA delivery was the first successful method used in producing fertile transgenic maize plants *(1)*. "Biolistic" is a generic term for microparticle bombardment and derives from biological ballistic *(2)*. Currently, microprojectile bombardment is among the most reliable and efficient direct DNA delivery systems for monocots. The original design of the biolistic gun used a gunpowder charge to propel the DNA-coated tungsten particles (microparticles) through a vacuum chamber into target cells *(3)*. The current biolistic device (Biolistics PDS-1000/He particle delivery system) uses DNA-coated gold particles driven by helium gas *(4)*.

According to BioRad PDS-1000/He manual (http://140.226.65.22/Davis_lab/Helminth_Protocols/PDS_1000_Manual.pdf), "The Biolistic PDS-1000/He system uses high pressure helium,

M. Paul Scott (ed.), *Methods in Molecular Biology: Transgenic Maize, vol. 526*

DOI: 10.1007/978-1-59745-494-0_3

released by a rupture disk, and partial vacuum to propel a macrocarrier sheet loaded with millions of microscopic tungsten or gold microcarriers toward target cells at high velocity. The microcarriers are coated with DNA or other biological material for transformation. The macrocarrier is halted after a short distance by a stopping screen. The DNA-coated microcarriers continue traveling toward the target to penetrate and transform the cells."

The velocity at which the "DNA bullets" (microcarriers) strike the target cells can be adjusted by *(1)* changing the helium pressure (psi of chosen rupture disk), *(2)* the amount of vacuum in the bombardment chamber, *(3)* the distance from the rupture disk to the macrocarrier, *(4)* the macrocarrier travel distance to the stopping screen, and *(5)* the distance between the stopping screen and target cells.

These parameter choices are critical to the success of gene delivery for recovering stable transgenic events. Typically, researchers use transient expression of a visual marker gene (such as the *gus* or *gfp* gene) for optimizing these conditions. Although transient assays can serve as an indicator of whether DNA delivery has occurred, they cannot be the only criteria used for optimization of stable transformation protocols.

Other particle acceleration methods based on similar concepts have also been developed. The ACCELL™ system uses an electrical impulse to accelerate the particles. The greater control of particle penetration by this device makes it a versatile tool for delivering DNA to a variety of crops, genotypes, and tissues *(5)*. The particle inflow gun (PIG) *(6, 7)* accelerates DNA-coated particles with a gentle burst of gas without the use of a macrocarrier, thus having the advantage of being less damaging to the target.

In establishing a protocol for maize transformation, the following factors should be considered: effectiveness of gene delivery methods, competence of targeted plant materials, and robustness of the selection systems used for recovery of transformed events. While the biolistic delivery method possesses obvious advantages for its ability to penetrate walled plant cells readily, thereby significantly expanding the range of transformation-competent plant tissues, its major disadvantage is that it generates a high frequency of multiple transgene insertion copies in the plant genome. Compared with low copy number insertion events, high copy number insertion events are prone to multisequence-induced gene silencing *(8)* and have been shown to cause lower stable transgene expression over generations *(9)*. On the other hand, because the biolistic delivery method offers greater flexibility in choice of plant target materials, it remains an efficient and powerful transformation method for many recalcitrant plant species.

The maize biolistic transformation method described here is based on Songstad et al.'s protocol of 1996 *(10)* and our previously published protocol in 2000 *(11)*. Our current average transformation frequency is 18% (transformation frequency is defined as the number of independent bialaphos-resistant calli per 100 bombarded embryos). Typically, bialaphos-resistant callus events can be identified 2–3 months after the day of bombardment experiment. Transgenic plantlets are regenerated 1–2 months later and seeds can be harvested in another 2–3 months. The total duration of this protocol (the day of bombardment to seed harvest) is about 8.5 months.

2. Materials

2.1. DNA Constructs

A typical DNA construct used in our biolistic-mediated maize transformation system contains the *bar* gene selectable marker cassette, which confers resistance to phosphinothrycin, the active ingredient in bialaphos *(12)*. We have used both the double 35S CaMV promoter *(13)* in construct pTF101.1 *(14)* and the maize ubiquitin promoter *(15)* in construct pBar184 *(11)* to drive the *bar* gene.

2.2. Plant Material

1. Maize Hi II F1 seeds (*see* **Note 1**): Ears of the maize Hi II genotype [A188 × B73 origin *(16)* harvested from green house grown embryo donor plants 9–12 days after pollination].
2. Maize B73 seeds (*see* **Note 1**): as pollen donor plant.

2.3. Stock Solutions

1. N6 Vitamin stock (1,000×): 1.0 g/L thiamine HCl, 0.5 g/L pyridoxine HCl, 0.5 g/L nicotinic acid, 2.0 g/L glycine. Store in 50 mL/L aliquots in Falcon tubes at –20°C. Thaw one tube at a time and store at 4°C.
2. MS Vitamin (modified) Stock (1,000×): 0.5 g/L thiamine HCl, 0.5 g/L pyridoxine HCl, 0.05 g/L nicotinic acid, 2.0 g/L glycine. Store 50 mL/L aliquots in Falcon tubes at –20°C. Thaw one tube at a time and store at 4°C.
3. 2,4-D, that is, (2,4-dichlorophenoxy acetic acid) stock: Weigh 0.25 g/L 2,4-D in a fume hood, and dissolve in 1 N KOH (10 mL/L) on low heat. When dissolved, bring up to 250 mL/L final volume with ddH_20 water. Store at 4°C in Duran bottle.
4. Silver nitrate stock: Dissolve 0.85 g/L silver nitrate (Fisher) in 100 mL/L of ddH_2O. Filter-sterilize the stock solution (50 mM) and store at 4°C for up to 1 year in a foil-wrapped container to avoid exposure to the light.

5. Bialaphos stock: Dissolve 100 mg of bialaphos (Duchefa Plant Biotechnology Products, *see* **Note 2**) in 100-mL/L ddH_2O. Filter-sterilized stock solution (1 mg/mL) is stored at 4°C for up to 4 months in 50-mL/L Falcon tubes.
6. Glufosinate stock: Dissolve 100 mg of glufosinate ammonia (Sigma, Cat. # D5417) in 100 mL of ddH_2O. Stock solution (1 mg/mL) is filter sterilized and stored in 50-mL Falcon tubes at 4°C.

2.4. Culture Media (see Notes 3 and 4)

1. Callus initiation (N6E) medium: 4 g/L N6 salts [Phytotech Labs; *(17)*], 1 mL/L N6 vitamin stock, 2 mL/L 2,4-D, 2.8 g/L l-proline, 30 g/L sucrose, 100 mg/L casein hydrolysate, 100 mg/L myo-inositol, 2.5 g/L gelrite; adjust pH to 5.8 using 1 N KOH. Silver nitrate stock (0.5 ml/L for a final concentration of 25 μM) is added after autoclaving.
2. Osmotic (N6OSM) medium: 4 g/L N6 salts, 1 mL/L N6 vitamin stock, 2 mL/L 2,4-D, 0.7 g/L L-proline, 30 g/L sucrose, 100 mg/L casein hydrolysate, 100 mg/L myo-inositol, 36.4 g/L sorbitol, 36.4 g/L mannitol, 2.5 g/L gelrite; adjust pH to 5.8 using 1 N KOH. Silver nitrate stock (0.5 ml/L for a final concentration of 25 μM) is added after autoclaving.
3. Selection (N6S) medium: 4 g/L N6 salts, 1 mL/L N6 vitamin stock, 2 mL/L 2,4-D, 100 mg/L myo-inositol, 30 g/L sucrose, 2.5 g/L gelrite; adjust pH to 5.8 using 1 N KOH. Bialaphos (2 mL/L) and silver nitrate stock (0.1 ml/L for a final concentration of 5 μM) are added after autoclaving.
4. Regeneration medium I: 4.3 g/L MS salts [Phytotech Labs; *(18)*], 1 mL/L modified MS vitamin stock, 100 mg/L myo-inositol, 0.25 mL/L 2,4-D, 30 g/L sucrose, 3 g/L gelrite; adjust pH to 5.8 using 1 N KOH. Bialaphos (2 ml/L) is added after autoclaving when medium is cooled
5. Regeneration medium II: 4.3 g/L MS salts (Phytotech Labs), 1 ml/L modified MS vitamin stock, 100 mg/L myo-inositol, 60 g/L sucrose, 3 g/L gelrite; adjust pH to 5.8 using 1 N KOH. Filter-sterilized glufosinate ammonia (6 ml/L) is added after autoclaving.
6. Regeneration medium III: 4.3 g/L MS salts, 1 mL/L modified MS vitamin stock, 100 mg/L myo-inositol, 30 g/L sucrose, 3 g/L gelrite; adjust pH to 5.8 using 1 N KOH.

2.5. Supplies for Biolistic Bombardment

1. 0.6-μm Gold particles (Bio-Rad, Hercules, CA).
2. Macrocarrier holders (Bio-Rad, Hercules, CA).
3. Macrocarriers (Bio-Rad, Hercules, CA).
4. 150-μm mesh screen (McMaster Carr, Elmhurst IL).

5. 100% Ethanol.
6. 2.5 M $CaCl_2$, filter-sterilized, 4°C.
7. 0.1 M Spermidine, free-base, filter-sterilized, aliquoted, and frozen at –20°C.
8. Ultrasonicator (Fisher, 18 oz. FS6 single channel sonic cleaner).
9. Vortex Genie 2 (Fisher).

2.6. Other Supplies and Reagents

1. Sterilizing solution: 50% commercial bleach (5.25% hypochlorite) and one drop of surfactant Tween 20 per liter.
2. Redi-Earth (Hummert, 4500 Earth City Expressway, Earth City, MO 63045, USA; Cat. # 10-2030-1).
3. Sunshine Universal Mix SB300 (Fosters Inc., P.O. Box 2674, Waterloo, IA 50704, USA).
4. Greenhouse flat with drainage holes (holds 32 small pots) (Hummert, Cat. # 11-3000-1).
5. Small pot (6.4 cm^2 each in four packs) (Hummert, Cat. # 11-0300-1).
6. Humi-dome (plastic, transparent) (Hummert, Cat. # 14-3850-1).
7. Large pot for Hi II (2-gal nursery pot with four drainage holes) (Hummert, Cat. # 14-9606-1).
8. Large pot for B73 pollen donor plants (3-gal nursery pot with four drainage holes) (Hummert, Cat. # 14-9612-1).
9. Osmocote Plus 16-8-12 (controlled release fertilizer tablets with trace elements) (Hummert, Cat. # 07-6450-1).
10. Calcium/magnesium solution (Dr. C. Block, USDA-ARS, North Central Regional Plant Introduction Station, Ames, IA, USA): First make two separate stock solutions as follows:
    (a) Stock #1: 720 g/L of $Ca(NO_3)_2{\cdot}4H_2O$.
    (b) Stock #2: 370 g/L of $MgSO_4{\cdot}7H_2O$ (Epsom salts).
    (c) To prepare working solution, add 5 mL of each Stock #1 and Stock #2 into 1 gal (~4 L) of H_2O (*see* **Note 5**).
11. Miracle Gro Excel 15-5-15 (water-soluble fertilizer supplemented with calcium and magnesium Hummert, Cat. # 07-5660-1).
12. Marathon® (restricted use pesticide for aphid control) (Hummert Cat. # 01-1118-1).
13. Whitemire yellow monitoring cards (for fungus gnat control) (Hummert Cat. # 01-2700-1).

14. Shoot bags (Lawson Bags, 318 Happ road, P.O. Box 8577, Northfield, IL 60093, Cat. # 217).
15. Striped (red or green) tassel bags (Lawson, Cat. # 404).
16. Plain (brown) tassel bags (Lawson, Cat. # 404).
17. Vent tape (1 in.) (Fisher, Cat. # 19-027-761).
18. Filter paper (Whatman No. 4, 5.5 cm) (Fisher, Cat. # 09-825-K).
19. Drierite (8 mesh indicating) (Fisher, Cat. # 07-578-3B).

3. Methods

3.1. Preparing Plant Materials for Biolistic Gun Transformation

3.1.1. Embryo Dissection

Both immature zygotic embryos (IZE) and IZE-derived embryogenic callus cultures from Hi II germplasm can be used for biolistic transformation.

1. Maize ears are harvested 9 (summer) to 12 (winter) days after pollination in the greenhouse. The ideal size of immature embryo for transformation should be between 1.2 and 2 mm (*see* **Note 6**).
2. Dehusk ear. Cut off and discard top 1 cm of ear and insert a straight-nosed forceps into the top end of the ear. This "handle" facilitates aseptic handling of the cob during embryo dissection. Place impaled ear and forceps into a sterilized mason jar in a laminar flow bench. If necessary, sterilize up to four ears in one mason jar.
3. Add ~700 mL of sterilizing solution to cover ear. During the 15–20 min disinfection, occasionally swirl the ears and tap the mason jar on the surface of the flow bench to dislodge air bubbles for thorough surface sterilization of ear. Holding on to the forceps, pour off bleach solution and rinse the ears three times in generous amounts of sterilized water. The final rinse is drained off and the ears are ready for embryo dissection (*see* **Notes 7** and **8**).
4. In a large (150 mm × 15 mm) sterile Petri plate, cut off the kernel crowns (the top 1–2 mm) with a sharp scalpel blade. Use sterilizing ovens for intermittent resterilizing of utensils throughout this protocol (*see* **Note 9**).
5. Excise the embryos by inserting the narrow end of a sharpened spatula between the endosperm and pericarp at the basipetal side of the kernel (toward the bottom of the cob) popping the endosperm out of the seed coat. This exposes the untouched embryo, which sits at the top side of the kernel, close to the kernel base. The embryo is

gently coaxed onto the spatula tip and plated with the embryo-axis side down (scutellum side up) onto a filter paper overlaying the N6E media in a 2 cm × 2 cm grid (30 embryos/plate).

6. Wrap the plate with a vent tape and incubate at 28°C in the dark for 2 or 3 days before bombardment, whichever is most convenient.
7. Alternatively, these embryos can also be kept in 28°C in the dark for 2 weeks to initiate Type II callus.

3.1.2. Type II Callus Initiation

1. After 2 weeks, friable, rapidly growing embryogenic callus can be seen proliferating from the embryo scutellar tissue of ~100% of the IZE explants. This tissue is subcultured to fresh N6E medium, and plates are wrapped with parafilm (28°C, dark).
2. Callus lines, each originating from an independent IZE source are developed for bombardment over a 6–8-week period by weekly subculturing of this material to fresh N6E medium (*see* **Note 10**).

3.2. Gold Particle Preparation

3.2.1. Washing Gold

1. Weigh 15-mg gold (0.6 μm) particles and transfer to sterile, 1.5 mL microcentrifuge tubes. These tubes are considered 10× gold quantities (*see* **Note 11**).
2. In the laminar flow hood, add 500 μL 100% ethanol, straight from the freezer, to each tube of 15-mg gold and sonicate in an ultrasonic water bath for 15 s. Tap the closed tube on the bench to gather all droplets to the tube bottom and let the tube sit until all the particles have settled out (up to 30 min).
3. Spin in a tabletop centrifuge for 60 s at 3,000 rpm, and remove the ethanol supernatant (keep teardrop-shaped pellet facing down or it will fall into the pipet).
4. To rinse, add 1 mL ice cold, sterile ddH_2O by dribbling the water down the side of the microcentrifuge tube. Slightly disturb the pellet by finger vortexing, and then let the gold settle out again.
5. Spin at 3,000 rpm for 60 s. Repeat the rinse step two more times, the third time centrifuging at 5,000 rpm for 15 s.
6. After removing the final wash, suspend the pellet in 500 μL sterile water.
7. Ultrasonicate this suspension for 15 s, then immediately place the tube on a vortex to keep it shaking as rapidly as possible (vortex setting of 3).
8. Leave the tube shaking at this setting while you open it and aliquot the washed gold for storage.

3.2.2. Aliquoting Gold to 1× Tubes

1. For each 10× tube of washed gold, set out ten, 1.5 mL microcentrifuge tubes in a microcentrifuge tube rack.
2. While the 10× tube is shaking, aliquot 250 μL of the gold suspension to each of the 10 tubes. Then, beginning with the last tube, start backward, aliquoting another 250 μL of the gold suspension to each tube.
3. When finished, each "1×" tube of gold should contain 1.5-mg gold in 50 μL water. Label the top of each tube as 1×, close, and freeze (–20°C) until use.

3.2.3. Coating the Gold with DNA

1. On bombardment day, leave the gold in the freezer until just before you begin the gold coating procedure. *Briefly*, thaw one, 1× tube of gold for 8–12 plates to be bombarded.
2. Ultrasonicate the tube for 15 s.
3. In the flow bench, add, in sequence, the appropriate quantities of selection construct and GOI construct (*see* **Note 12**), finger-vortexing the tube after each.
4. Finger-vortex the tube well and then tap it on the bench top to gather all the droplets to the bottom of the tube.
5. Add 50 μL $CaCl_2$. Using the same pipet tip, gently suck the suspension up and down once, then place the tube on the vortex at low speed (setting 2–3 shaking on Vortex Genie 2).
6. Add 20 μL spermidine while the tube is still shaking on the vortex. Wait 30 s, close the tube, and finger-vortex well. Return the tube to the vortex and let it shake for 10 min.
7. Remove the tube from the vortex and let the gold settle out for several minutes.
8. Centrifuge for 15 s at 5,000 rpm (just enough to pellet the gold), and then pipet off the supernatant.
9. Take the 100% ethanol from the freezer (*see* **Note 13**) and add 250 μL cold ethanol to the 1× pellet.
10. Finger-vortex to dislodge the pellet, and then rock the tube back and forth until the gold achieves a very "silty" smooth consistency dispersed on the base of the tube.
11. Let the tubes sit until the gold settles out (3–5 min). Centrifuge for 15 s at 5,000 rpm. Remove the supernatant and add 140 μL 100% ethanol.
12. Finger-vortex well (do not sonicate) to ensure complete suspension of the gold pellet and place on the vortex (setting 2–3) for loading the macrocarriers.

3.2.4. Loading the Macro-carriers

1. Fit presterilized macrocarriers (*see* **Note 14**) into their stainless steel holders in a sterile dish surrounded by indicating Drierite (*see* **Note 15**).
2. While the 1× tube of DNA-coated gold particles is still shaking, open the tube and aliquot 10 μL of the suspension onto the center of each macrocarrier (*see* **Note 16**).
3. After loading the macros, let them sit for 5–10 min to be sure that they are dry before bombardment.

3.3. Bombardment

3.3.1. Embryo Bombardment

1. After 2–3 days of preculture on N6E, the embryos will be ridged and swollen as Type II callus initiation has begun. This is an appropriate stage for bombardment.
2. Draw a 3.5-cm diameter circle on the bottom of a plate of N6OSM medium.
3. Four hours prior to bombardment, use sterile forceps to transfer the embryos and filter paper onto the N6OSM medium (*see* **Note 17**). Embryos should be facing scutellum side up at bombardment since it is from this surface that subsequent callus initiation begins from which transformed cells are then selected.
4. Load 650-psi rupture disk.
5. Assemble the macrocarrier launch assembly by first laying in place a stopping screen followed by an inverted, preloaded macrocarrier holder (*see* **steps 1–3** in **Subheading 3.2.4**), which is held in place by screwing on the launch assembly lid.
6. Slide the launch assembly into place immediately below the helium nozzle. Pre-set gap distance to 6 mm.
7. Slide a sterilized 150-μm mesh screen onto the shelf directly below the launch assembly. This screen is supported on a second plexiglass stage (like the one that holds the Petri plate at bombardment) with a 3.8-cm-diameter hole cut in the middle of it (*see* **Note 18**).
8. Slide the opened Petri dish containing the target tissue onto the shelf at a selected distance from the stopping screen (6 cm).
9. The vacuum chamber is closed, a vacuum pulled, and the gun fired in time for the rupture disk to break as soon as the vacuum reaches 28 in. of Hg.
10. The chamber is vented, the plate containing the bombarded tissue is removed, and the gun is prepared for the next bombardment by replacing the spent rupture disk, macrocarrier, and stopping screen (disposables). All plasmid waste is disposed of in biohazard bags for autoclaving.

11. Repeat **steps 4–10** for each shot.
12. The bombarded embryos (still on filter paper on N6OSM) are gently wrapped with vent tape and incubated in 28°C in the dark.
13. The next day (16–20 h after bombardment), embryos are transferred off the filter paper on N6OSM to the surface of N6E media with no filter paper to continue callus initiation. Embryos are again oriented scutellum side up and plates are wrapped with vent tape.

3.3.2. Callus Bombardment

1. Typically, 6–8 weeks after callus initiation, friable and rapidly growing embryogenic Type II callus derived from the embryo scutellar tissue can be ready for bombardment.
2. Draw a 3.5-cm-diameter circle on the bottom of a plate of N6OSM medium. This defines the target area to which callus pieces are loaded for bombardment.
3. Four hours prior to the bombardment, use a microscope to transfer 30 callus pieces (4 mm) from a friable, rapidly growing callus line to the target area.
4. Bombardment parameters for Hi II callus are identical to those used for Hi II IZE (refer to **steps 4–11** in **Subheading 3.3.1** above).
5. The bombarded callus is left on N6OSM medium for 1 h after bombardment and then transferred to N6E medium maintaining the integrity of each bombarded piece.
6. Plates are wrapped with vent tape (28°C, dark).

3.4. Selection for Stable Transgenic Events from Bombarded Embryos or Callus

1. After 7 days on initiation medium (N6E) the bombarded embryos or callus pieces are transferred to N6S selection medium (2.0 mg/L bialaphos) to begin the recovery of transformed cells.
2. Plates are wrapped with parafilm throughout selection, and all embryo explants or callus pieces are subcultured intact throughout selection (*see* **Note 19**).
3. Two weeks later, embryos or callus pieces are transferred to fresh N6S. This step is repeated once more before bialaphos-resistant events are visible emerging from selected immature embryos or callus pieces (within 6–8 weeks from date of bombardment). Each of these proliferating callus clumps is considered an independent putative transgenic event (*see* **Note 20**).

3.5. Regeneration of Hi II Transgenic Plants

1. Regeneration of transgenic Type II callus (friable, stocked somatic embryos present) is accomplished by subculturing

about 12 pieces (10-mm diameter) of embryogenic callus (use a 40× dissecting scope) to regeneration medium I.

2. Plates are wrapped with vent tape and incubated (dark, 25°C) for 10 to 14 days after which 15 small pieces of callus (approximately 4 mm) enriched with stocked somatic embryos are subcultured (again using the scope) from callus on regeneration medium I to regeneration medium II.
3. These plates are incubated for 2–3 weeks at 25°C in the dark (*see* **Note 21**). Petri plates are wrapped with vent tape.
4. After this maturation period (2–3 weeks), many of the somatic embryos on regeneration medium II are opaque and white with a clearly defined scutellar region. From some, the coleoptile is already emerging. Again using a dissecting scope, transfer these mature somatic embryos (~12 pieces per plate) to the surface of regeneration medium III for germination in the light (25°C, 80–100 $\mu E/m^2/s$ light intensity, 16:8 photoperiod). Again wrap the plates with vent tape.
5. Each somatic embryo germinates to form leaves and roots on regeneration medium III within 1 week and plantlets are ready for transfer to soil within 10 days (*see* **Note 22**).

3.6. Transplanting and Acclimation

1. In a laminar flow hood, use sterile forceps to transfer plantlets (a good-sized plantlet measures about ~5 cm) from the Petri plate to the soil surface of a small pot filled with Redi-Earth (*see* **Note 23**).
2. Remove any medium still clinging to the roots. Plants should be handled with extreme care to avoid breaking off the leaf.
3. Plantlet roots are gently pressed into the soil and covered. Place small pots into greenhouse flat with drainage holes. Thoroughly soak the flat with a gentle stream of water so as not to dislodge the transplants.
4. Place the flat in the growth chamber and cover it with a humi-dome in which one ventilation hole has been cut (*see* **Note 24**).
5. Flats should not need water for 48 h if thorough soaking was done at transplant. After that, water individual plants only as needed.
6. Remove the humi-dome when the plants are tall enough to touch it, and 1 week later move the flat from the growth chamber to the greenhouse (*see* **Note 25**).

3.7. Greenhouse Care of Transgenic Plants

1. Once transgenic plants have been moved to the greenhouse continue to monitor soil moisture on a per plant basis. Water only if dry, using a watering can with a well-defined spout.

2. Transgenic plantlets are fertilized once using liquid Miracle Gro Excel 15-5-15 (20 ppm), a low concentration water-soluble fertilizer, before transplanting to big pots (*see* **Note 26**).
3. At transplant to big pots, add one (15 g per 2 gallon pot) Osmocote Plus 16-8-12 tablet to bottom half of soil profile. At 6-7 leaf stage, add an additional tablet to soil surface of each pot.
4. After transplant to big pots, plants are watered three times with Miracle Gro Excel 15-5-15 (50-100 ppm). Following this, calcium/magnesium solution is given to all plants once every 2 weeks.
5. We cross all our R_0 female transgenic plants by pollinating them with nontransgenic donor pollen.
6. To provide nontransgenic donor pollen to pollinate transgenic ears, begin by planting 2 donor seeds twice per week (4 seeds per week) as soon as the first transgenic material is transferred to the light on regenerate medium I.
7. Tassels of all R_0 transgenic plants are bagged as soon as tassels emerge to minimize transgenic pollen flow in the greenhouse. In addition, transgenic plants are grown in a separate room from the nontransgenic, pollen donor plants.
8. After pollination, watering is continued as needed until 21–25 days later at which time watering is stopped altogether, and plants are moved to a dry-down area.
9. To further aid in cob dry-down, lift the pollination bag off the ear 15 days postpollination. Ten to 15 days later pull down the husks to facilitate further drying of the kernels (*see* **Notes 27** and **28**).
10. Forty to 45 days after pollination, harvest the seed (*see* **Note 29**). Seed is inventoried and securely stored in the cold (0–3°C, 60% relative humidity).

4. Notes

1. We typically plant Hi II F_1 seeds year round in the greenhouse and obtain F_2 immature embryos (by self- or sib-pollination of F_1 plants) for transformation experiments. F_1 seeds are produced in the field by pollinating Hi II parent A silks with Hi II parent B pollen. These parent lines and B73 can be obtained from the Maize Genetics Cooperation Stock Center (https://maizecoop.cropsci.uiuc.edu/request/) and are increased and

maintained in our field by sib-pollination when feasible. Hi II and B73 plants generally take 60 and 70 days (depending on the season), respectively, to flower in the greenhouse.

Corn plants in our greenhouse are placed in pots on the ground beginning 2 weeks after transplant to large pots. Our greenhouse operates on a 16:8 photoperiod. The average temperature is 28°C (day) and 21°C (night). The light intensity (230 μE/m^2/s at 3.5 ft above ground) was measured in February on a slightly overcast day, therefore, does not factor in any additional sunlight.

To provide a steady flow of immature embryos for transformation experiments, 10 Hi II F_1 seeds are planted twice per week to guarantee 15 ears per week to the lab.

2. Gold Bio Technology Inc. (Duchefa Plant Biotechnology Products), St. Louis, MO, USA at http://www.goldbio.com, Cat. # B0178.
3. Media described in **steps 1–3** in **Subheading 2.4** use 100 × 15 Petri plates (Fisher) and are poured at a volume of 40 plates per liter. These media are derived from Songstad et al. (19) and Vain et al. (20) (for step 2 in Subheading 2.4).
4. Regeneration media described in steps 4–6 in **Subheading 2.4** use 100 × 25 Petri plates (Fisher) and are poured at a volume of 32 plates per liter. Regeneration medium II is after Armstrong and Green (21) and McCain et al. (22). All media are dried thoroughly before storage at room temperature in the dark.
5. The stock solutions should be made separately rather than adding both salts to one bottle of water. If they are not made separately, gypsum will immediately be formed.
6. Immature ears harvested from the greenhouse (or field) are stored in their husks and pollination bags in a larger plastic bag at 4°C. If there are not enough freshly harvested ears for an experiment, ears stored from Friday through Sunday, or Tuesday through Thursday, are used for experiments on Monday and Friday, respectively. We have not experimented with ears stored for longer than 5 days.
7. For surface sterilizing a large number of ears, we use a preautoclaved 4-L beaker that will hold up to 20 ears at a time.
8. Greenhouse ears can be surface sterilized for as little as 15 min, but we always use the full 20 min when sterilizing field ears from which we routinely encounter more problems with fungal or bacterial contamination in experiments.
9. Intermittent resterilization of all utensils used for dissection is accomplished using a Steriguard 350 bead sterilizer (Inotech Biosystems International, Rockville, MD, USA).

10. Callus lines are discarded 4 months after initiation to minimize detrimental effects (such as reduced regenerability and poor fertility) that may result from extended time in tissue culture.
11. At least two tubes of gold are always washed at one time so that they balance each other in the centrifuge steps.
12. We currently use 0.03 mg of selectable marker per 1× gold tube + threefold more GOI, adjusted for molar equivalents depending on relative sizes of the plasmids.
13. Ethanol (100%) is properly sealed and stored in the freezer when not in use.
14. We sterilize macrocarriers by soaking 10 min in 70% ethanol and then air-dry over night.
15. After loading, macrocarriers are laid over a bed of Drierite to maintain dryness during bombardment.
16. While aliquoting the suspension, draw a half-spiral with the pipet tip from the center and outward of each macrocarrier to ensure the even distribution of the suspension over the inner, target circle. *Important!* Work quickly to avoid evaporation of the remaining suspension.
17. Center the embryo grid, not the filter paper, on the 3.5-cm-diameter circle drawn on the bottom center of the plate.
18. The mesh screen is sterilized by autoclaving. We typically reuse this screen for 8–12 shots (with same DNA construct) before discarding.
19. Both bombarded IZE and callus pieces are transferred intact throughout selection – there is no need to break proliferating callus into pieces during selection. This saves time and confusion about which piece derives from which explant.
20. Using the described protocol, at an average frequency of 18%, we expect to recover eighteen independent, bialaphos-resistant calli per 100 bombarded embryos. When using Hi II germplasm in which we recover few to no escapes (events that do not carry the *bar* selectable marker gene), we calculate transformation frequency as (number of bialaphos-resistant calli recovered ÷ total number of embryos bombarded) × 100.
21. Glufosinate ammonium contains the same active ingredient (phosphinothrycin) as bialaphos. It is used instead of bialaphos during this regeneration step to save cost. In our experience, glufosinate must be used at higher concentrations than bialaphos to achieve the same effect. Imposing continued selection pressure during this regeneration step (whether glufosinate or bialaphos) is effective because non-transgenic callus does not form mature somatic embryos on

this medium. As such, only callus containing the *bar* gene is advanced to the light after this final in vitro selection step.

22. Using this protocol we expect an average of 8 out of 10 callus events to produce healthy plantlets.
23. Typically, plantlets on each plate germinate at different rates. After transplanting the large plantlets, smaller plantlets in the same Petri plate are returned to the light until they too are large enough for transplant to soil.
24. A Conviron (Controlled Environments Limited, 590 Berry St. Winnipeg, MB, Canada) growth chamber is used for this intermediate step. The conditions are 16:8 photoperiod; 350 $\mu E/m^2/s$ light intensity (plant height of 30 cm) with a combination of fluorescent and incandescent bulbs; 26°C (day) and 22°C (night).
25. If the greenhouse is in a different building than the laboratory and the outdoor temperature is below freezing or very cold, special care is needed during this step. Cover the flat with a humi-dome, wrap it in a plastic garbage bag, and transport it to the greenhouse in a preheated vehicle.
26. A plantlet is ready to be transplanted to a big pot if soil adheres to its root ball when lifted out of the small pot. Plantlets are generally over 16-cm tall at this stage.
27. If seeds are contaminated by fungus, use 70% ethanol to clean the surface before storage.
28. Common pests found in our greenhouse are aphids (on the tassels or on the underside of leaves), spider mites (on the underside of leaves), and thrips (on the leaf whorl). We spray for mites or thrips once a month, or as needed, using Floramite®, Pylon®, Akari®, Avid®, or Samite® for spider mite control and Conserve® for thrip control (all from Hummert). Fungus gnats are another common pest and often become a problem when plants are overwatered. Place yellow monitoring cards (to which the air-borne gnats stick) around the greenhouse to control them. Another disease that we see although infrequently is smut. Smutted plants are immediately discarded. To limit disease onset, the greenhouse must be kept clean. The floor should be frequently swept or sprayed clean; the drain hole should always be left unclogged, and fallen and dead leaves still clinging to the plants should be removed. Garbage should be emptied once a week. Mice moving in from the field in the fall will feed on maize seed (transgenic or otherwise). To solve this problem, we set out mice traps bated with peanut butter each autumn.
29. On an average, 70% of regenerated events will produce >50 kernels (we take 3–4 plants per event to the greenhouse). Because we have observed consistently poor transgenic seed

sets for plants flowering in the greenhouse in July (in Ames, IA) we now avoid February transformations for constructs going to seed.

Acknowledgments

The authors wish to thank current and former maize team and greenhouse colleagues, Tina Paque, Marcy Main, Jennie Lund, Jennifer McMurray, and Lise Marcell for their contributions to the development of this protocol. This research is supported partially by the National Science Foundation (DBI #0110023), the Iowa State University Agricultural Experiment Station, the Office of Biotechnology, the Plant Science Institute, and the Baker Endowment Advisory Council for Excellence in Agronomy at Iowa State University.

References

1. Gordon-Kamm, W.J., Spencer, T.M., Mangano, M.L., Adams, T.R., Daines, R.J., Start, W.G., O'Brien, J.V., Chambers, S.A., Whitney, J., Adams, R., Willetts, N.G., Rice, T.B., Mackey, C.J., Krueger, R.W., Kausch, A.P., and Lemaux, P.G. (1990) Transformation of maize cells and regeneration of fertile transgenic plants. *Plant Cell* 2, 603–618.
2. Armstrong, C.L. (1999) The first decade of maize transformation: a review and future perspective. *Maydica* 44, 101–109.
3. Klein, T.M., Wolf, E.D., Wu, R., and Sanford, J.C. (1987) High-velocity microprojectiles for delivering nucleic acids into living cells. *Nature* 327, 70–73.
4. Sanford, J.C., Devit, M.J., Russell, J.A., Smith, F.D., Harpending, P.R., Roy, M.K., and Johnston, S.A. (1991) An improved, helium-driven biolistic device. *Tech J Meth Cell Mol Biol* 3, 3–16.
5. McCabe, D. and Christou, P. (1993) Direct DNA transfer using electric discharge particle acceleration (ACCELL™ technology). *Plant Cell Tissue Organ Cult* 33, 227–236.
6. Finer, J.J., Vain, P., Jones, M.W., and McMullen, M.D. (1992) Development of the particle inflow gun for DNA delivery to plant cells. *Plant Cell Rep* 11, 323–328.
7. Takeuchi, Y., Dotson, M., and Keen, N.T. (1992) Plant transformation: a simple particle bombardment device based on flowing helium. *Plant Mol Biol* 18, 835–839.
8. Matzke, M.A., Aufsatz, W., Kanno, T., Mette, M.F., and Matzke, A.J. (2002) Homology-dependent gene silencing and host defense in plants. *Adv Genet* 46, 235–275.
9. Shou, H., Frame, B., Whitham, S., and Wang, K. (2004) Assessment of transgenic maize events produced by particle bombardment or *Agrobacterium*-mediated transformation. *Mol Breed* 13, 201–208.
10. Songstad, D.D., Armstrong, C.L., Peterson, W.L., Hairston, B., and Hinchee, M.A.W. (1996) Production of transgenic maize plants and progeny by bombardment of Hi II immature embryos. *In Vitro Cell Dev Biol Plant* 32, 179–183.
11. Frame, B.R., Zhang, H., Cocciolone, S.M., Sidorenko, L.V., Dietrich, C.R., Pegg, S.E., Zhen, S., Schnable, P.S., and Wang, K. (2000) Production of transgenic maize from bombarded Type II callus: effect of gold particle size and callus morphology on transformation efficiency. *In Vitro Cell Dev Biol Plant* 36, 21–29.
12. White, J., Chang, S.-Y.P., Bibb, M.J., and Bibb, M.J. (1990) A cassette containing the bar gene of *Streptomyces hygroscopicus*: a selectable marker for plant transformation. *Nucleic Acids Res* 18, 1062.
13. Odell, J.T., Nagy, F., and Chua, N.H. (1985) Identification of DNA sequences required for activity of the cauliflower mosaic virus 35S promoter. *Nature* 313, 810–812.

14. Paz, M., Shou, H., Guo, Z., Zhang, Z., Banerjee, A., and Wang, K. (2004) Assessment of conditions affecting Agrobacterium-mediated soybean transformation using the cotyledonary node explant. *Euphytica* 136, 167–179.
15. Christensen, A.H. and Quail, P.H. (1996) Ubiquitin promoter-based vectors from high-level expression of selectable and/or screenable marker genes in monocotylendonous plants. *Transgenic Res* 5, 213–218.
16. Armstrong, C.L., Green, C.E., and Phillips, R.L. (1991) Development and availability of germplasm with high Type II culture formation response. *Maize Genet Coop News Lett* 65, 92–93.
17. Chu, C.C., Wang, C.C., Sun, C.S., Hsu, C., Yin, K.C., Chu, C.Y., and Bi, F.Y. (1975) Establishment of an efficient medium for anther culture of rice through comparative experiments on the nitrogen source. *Sci Sin* 18, 659–668.
18. Murashige, T. and Skoog, F. (1962) A revised medium for rapid growth and bio-assays with tobacco tissue cultures. *Physiol Plant* 15, 473–497.
19. Songstad, D.D., Armstrong, C.L., and Petersen, W.L. (1991) $AgNO_3$ increases type II callus production from immature embryos of maize inbred B73 and its derivatives. *Plant Cell Rep* 9, 699–702.
20. Vain, P., McMullen, M.D., and Finer, J.J. (1993) Osmotic treatment enhances particle bombardment-mediated transient and stable transformation of maize. *Plant Cell Rep* 12, 84–88.
21. Armstrong, C.L. and Green, C.E. (1985) Establishment and maintenance of friable, embryogenic maize callus and the involvement of l-proline. *Planta* 164, 207–214.
22. McCain, J.W., Kamo, K.K., and Hodges, T.K. (1988) Characterization of somatic embryo development and plant regeneration from friable maize callus cultures. *Bot Gaz* 149, 16–20.

Chapter 4

Agrobacterium-Mediated Maize Transformation: Immature Embryos *Versus* Callus

Vladimir Sidorov and David Duncan

Summary

Transformation with *Agrobacterium tumefaciens* is the preferred method for delivery of transgenes into a wide range of plant species including maize. Optimized protocols for the *Agrobacterium*-mediated transformation of freshly isolated immature embryos and embryogenic Type I callus derived from plant seedlings are described. These protocols are suitable for the transformation of a wide variety of corn genotypes including commercial inbred lines. *Agrobacterium* harboring a binary vector containing the neomycin phosphotransferase (*npt*II) or the glyphosate resistant 5-enolpyruvylshikimate-3-phosphate (EPSPS) as selectable marker genes and also the green fluorescence protein gene (*gfp*) have been used. GFP is a visual screening marker which allows tracking of transformation during different selection and regeneration steps. The described protocols provide double digit transformation frequencies and can be routinely used for the production of a large numbers of transgenic plants.

Keywords: *Agrobacterium tumefaciens*, Transformation, Maize, Immature embryo, Embryogenic callus

1. Introduction

Agrobacterium tumefaciens-mediated transformation is the most widely used method for genetic modification of agronomically important crops including corn. The fast progress in transformation of monocotyledonous species mediated by *Agrobacterium* was due to the development of an efficient transformation method for rice described by Hiei and co-authors *(1)* from Japan Tobacco Inc. A similar approach, based on disarmed *Agrobacterium* carrying the "super binary" vector with selectable marker genes, was used to establish a corn transformation protocol using

M. Paul Scott (ed.), *Methods in Molecular Biology: Transgenic Maize, vol. 526*

DOI: 10.1007/978-1-59745-494-0_4

freshly isolated immature embryos (IE) as initial explants *(2–4)*. More recently, an efficient transformation system with an average transformation frequency of 5.5 transformed plants per 100 inoculated IE was established for *Agrobacterium*-mediated transformation of maize IE using a standard binary vector system *(5)*. After considerable improvement of the system it has become a routine practice for efficient production of transgenic corn.

In most cases freshly isolated or pre-cultured IE are used for transformation.

Embryogenic callus, particularly Type I callus derived from IE, is another suitable material for *Agrobacterium* transformation. Optimization of transformation protocols, particularly with the desiccation of callus tissue during co-culture with *Agrobacterium*, has facilitated using callus also for transformation *(6–8)*. Recently it was demonstrated that embryogenic callus can be obtained with high efficiency from seedlings of mature seeds and transformed in a similar way as IE-derived callus *(9)*. A transformation system based on using seedling-derived callus does not require IE. *Agrobacterium*-mediated transformation of seedling-derived callus can be used as an alternative for conventional transformation of IE or IE-derived callus.

2. Materials

Standard laboratory safety procedures in addition to any requirements indicated for the material (MSDS, equipment manual, worker safety protocol, etc.) being used in this procedure:

1. Fungicide-treated corn seeds. H99 or other genotypes with high *in vitro* response is recommended.
2. Field or greenhouse-grown ears harvested 10–13 days post-pollination with embryos in a size range of 1.5–1.8-mm long. Genotypes with high *in vitro* response are recommended.
3. 90 × 25 mm Petri dishes.
4. Phytatrays™ or other culture vessels.
5. 200-ml beaker.
6. 250-ml flasks.
7. Plastic pipette.
8. Fume hood with vacuum capability or M2 biohazard safety cabinet.
9. Horizontal or vertical laminar flow hood.
10. Concentrated HCl.
11. Sodium hypochlorite (6.15% active ingredient).

12. Vacuum desiccator.
13. Rubber bands.
14. Parafilm M®.
15. Whatman #1 filter paper.
16. Translucent plastic box.
17. Forceps, scalpels, and spatula.
18. Bead sterilizer.
19. Regular dissecting microscope (Zeiss or other brand).
20. Dissecting microscope equipped with a GFP filters (can be Leica MZ80 with filter set (HQ480/40 excitation filter and HQ535/50 emission filter).
21. Table top centrifuge.
22. Spectrophotometer.
23. Rotary shaker 150 rpm.
24. Dark, 24°C growth culture chamber.
25. Lighted, 28°C growth culture chamber.
26. Dark, 28°C growth culture chamber or warm room.
27. Media.

All semi-solid media were solidified with 3.0 g/l Phytagel™ (Sigma, *see* **Note 1**). Exceptions were the media with paromomycin (Par) and silver nitrate, which had 7 g/l Gibco Phytagar™ (Invitrogen, *see* **Note 1**). Other gelling agents used are mentioned within the text. In all media the ascorbic acid, antibiotics, glyphosate, acetosyringone, and silver nitrate were filter sterilized. The pH of all media, if not specified, was adjusted to 5.7 or 5.8 before autoclaving. All ingredients in media if otherwise not specified were from Sigma (*see* **Note 1**).

(a) LBssck: 20.6 g/l LB Broth EZMix (Sigma, *see* **Note 1**). Before using, add 100 mg/l kanamycin, 50 mg/l spectinomycin, 50 mg/l streptomycin and 25 mg/l chloramphenicol (*see* **Note 2**). Semi-solid medium contains 15 g/l Bacto agar, pH 5.4 (*see* **Note 3**).

(b) AB/50 km, 25 spec, 200 AS, induction medium *(10)*: 1 g/l NH_4Cl, 0.3 g/l $MgSO_4 \cdot 7H_2O$, 150 mg/l KCl, 10 mg/l $CaCl_2$, 2.5 mg/l $FeSO_4 \cdot 7H_2O$, 1 ml/l NaH_2PO_4 (0.5 M stock), 20 g/l glucose, 19.52 g/l MES, pH 5.4. Before using for *Agrobacterium* induction add 200 μM acetosyringone, 50 mg/l kanamycin, and 25 mg/l spectinomycin.

(c) ½ MS PL, inoculation medium: 2.16 g/l MS *(11)* basal salt mixture (Phytotechnology Labs, *see* **Note 1**), 10 ml/l MS vitamins 100× (Phytotechnology Labs, *see* **Note 1**), 68.5 g/l sucrose (Phytotechnology Labs, *see* **Note 1**), 36 g/l glucose (Phytotechnology Labs, *see* **Note 1**), 0.115 g/l L-proline.

(d) ½ MS, co-culture medium: 2.16 g/l MS basal salt mixture (Phytotechnology Labs, *see* **Note 1**), 10 ml/l MS vitamins 100× (Phytotechnology Labs, *see* **Note 1**), 1 ml/l thiamine-HCl (0.5 mg/ml stock), 3 ml/l 2,4-D (1 mg/ml stock), 10 g/l glucose (Phytotechnology Labs, *see* **Note 1**), 20 g/l sucrose (Phytotechnology Labs, *see* **Note 1**), 0.115 g/l L-proline, 0.2 ml/l acetosyringone (1 M stock), 1.7 ml/l silver nitrate (2 mg/ml stock).

(e) MSVS34, germination medium: 4.33 g/l MS basal salt mixture (Phytotechnology Labs, *see* **Note 1**), 10 ml/l MS vitamins 100x, 40 g/l maltose, 0.5 g/l glutamine, 0.1 g/l casein hydrolysate, 0.75 g/l magnesium chloride, 1.95 g/l MES, 7.0 g phytagar, 6.0 ml/l 6-benzyladenine (0.5 mg/ml), 10.0 ml picloram (1 mg/ml), 2 ml/l ascorbic acid (50 mg/ml).

(f) MSW57, callus induction and maintenance medium: 4.33 g/l MS basal salt mixture (Phytotechnology Labs, *see* **Note 1**), 10 ml/l MS vitamins 100×, 1.25 ml thiamine HCl (0.4 mg/ml), 30 g/l sucrose, 1.38 g/l l-proline, 0.5 g/l casamino acids (Difco® *see* **Note 1**), 3.0 g/l phytagel, 0.5 ml/l 2,4-D (1 mg/ml), 2.2 ml/l picloram (1 mg/ml), and 1.7 ml/l silver nitrate (2 mg/ml).

(g) MSW57/0.1 gly/C500 and MSW57/0.25 gly/C500, selection media: MSW57 medium containing with 2 ml/l carbenicilin (250 mg/ml) and 0.2 ml/l or 0.5 ml/l glyphosate (0.5 M stock).

(h) MSBA, regeneration medium: 4.33 g/l MS basal salt mixture (Phytotechnology Labs, *see* **Note 1**), 1 ml/l Fromm vitamins *(12)*, 1,000× stock, 7.04 ml/l 6-benzyladenine (0.5 mg/ml), 30 g/l sucrose, 1.36 g/l L-proline, 0.05 g/l casamino acids (DifCo), 1 ml/l carbenicilin (250 mg/ml), and 0.2ml/l glyphosate (0.5 M stock) or 100 mg/l paromomycin).

(i) MSGR$^-$, regeneration/rooting medium: 4.33 g/l MS basal salt mixture (Phytotechnology Labs, *see* **Note 1**), 10 ml/l Fromm vitamins 1,000×, 10 g/l glucose (Phytotechnology Labs, *see* **Note 1**), 20 g/l maltose (Phytotechnology Labs, *see* **Note 1**), 0.15 g/l asparagine monohydrate, 0.1 g/l myo-inositol, 1 ml/l carbenicilin (250 mg/ml stock), and 0.2 ml/l glyphosate (0.5 M stock) or 100 mg/l paromomycin.

3. Methods

3.1. Transformation of Immature Embryos

Different steps of the transformation process are shown in **Fig. 1**. The GFP was used as a visual marker during selection and regeneration stages. This procedure, when followed, should result in

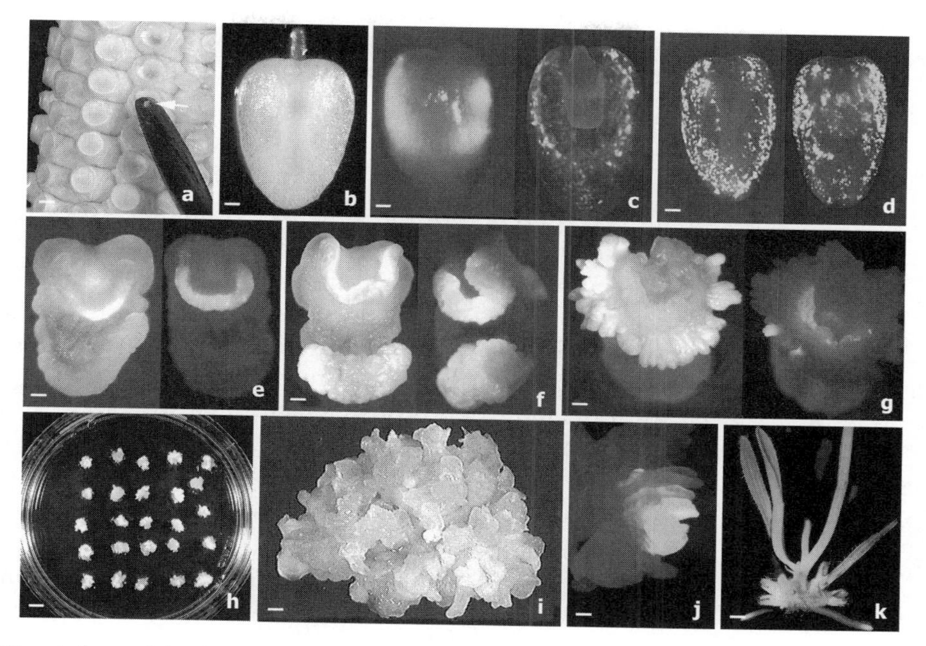

Fig. 1. Different stages of *Agrobacterium*-mediated transformation of corn using freshly isolated immature embryos (IE). (**a**) Ear of corn with tops of kernels removed. *Arrow* indicates isolated IE at the end of spatula (bar = 3 mm). (**b**) Freshly isolated IE with scutellum surface up (bar = 0.16 mm). (**c**) Transient GFP expression in IE (scutellum up at *left*, and scutellum down at *right*; first day on selection medium), which were placed on medium during co-cultivation with scutellum side facing down. This is not correct orientation of IE during co-culture. These IE usually do not produce transgenic callus and plants. No GFP positive spots can be seen on surface of scutellum which forms embryogenic callus (bar = 0.18 mm). (**d**) Two IE at the first day on selection medium which had transient GFP expression on the upper surface of scutellum. During the co-culture with *Agrobacterium* the scutellum of embryos was facing up (bar = 0.2 mm). (**e**) IE 7 days on selection. IE were on selection medium with scutellum side up. To the *left* is IE in incandescent light, and at the *right* is IE at the *blue* light. Transient GFP expression is still visible at the area of callus formation (bar = 0.3 mm). (**f**) The same as on (**e**). Callus is forming at both side of IE (bar = 0.4 mm). (**g**) IE after 12 days on selection medium. To the *left* is IE in the incandescent light, and at the *right* is IE at the *blue* light. Callus area with stable GFP expression is visible (bar = 0.5 mm). (**h**) Individual calli derived from IE after 1 month of selection (bar = 6 mm). (**i**) Higher magnification of single callus line from plate shown on (**h**) (bar = 0.4 mm). (**j**) Stable expression of GFP in callus observed after 1 month of selection (bar = 0.3 mm). (**k**) GFP expression in regenerated plants obtained 3 months after inoculation (bar = 2 mm) (*see Color Plates*).

an average transformation frequency (total number of events producing plants/number of embryos put into selection) of approximately 15%.

3.1.1. Agrobacterium Preparation

1. Streak out *Agrobacterium* from a glycerol stock on to solid LBsskc medium. Seal the LBsskc plate with Parafilm M® and place the plate upside down in a 28°C incubator for 2 days (*see* **Note 4**).
2. Move the plate with *Agrobacterium* to 2–8°C refrigerator (*see* **Note 5**).
3. In the late afternoon 2 days before inoculating the IE, pick one colony of the *Agrobacterium* from the plate and inoculate 25 ml of liquid LBsk (LB with 100 mg/l each of spectinomycin

and kanamycin) in a 250-ml flask. Place the flask on a shaker at approximately 150 rpm in the dark and 26°C overnight.

4. The next morning make a 1:5 dilution of the *Agrobacterium* culture in LBsk (add 10 ml of the culture into 40 ml fresh LBsk).
5. In late afternoon of the same day, divide the *Agrobacterium* culture into two 50-ml test tubes and centrifuge at 3,500 rpm for 15 min. Remove the supernatant and resuspend the cells in 10 ml induction broth with 200 μM of acetosyringone and 50 mg/l each of spectinomycin and kanamycin. Check the density of the suspension culture and dilute it to $O.D._{660} = 0.2$ with the induction broth supplemented with acetosyringone, spectinomycin, and kanamycin. The final volume is 50 ml in a 250-ml flask. Place the flask on a shaker at approximately 150 rpm in the dark and 26°C overnight.
6. The next morning, centrifuge the *Agrobacterium* cells and resuspend the pellet in 6–10 ml of inoculation medium with 200 μM acetosyringone. Check the O.D. at 660 and adjust it to an $O.D._{660}$ of 1.0.

3.1.2. Isolation of IE

1. Carefully remove the husks and silk from ears which are harvested 10–13 days post-pollination. Insert a blunt tip holder (can be forceps) at the basal end of ear (*see* **Note 6**).
2. Surface-sterilize ears for 20 min with a 50% dilution (v:v) of a 6.15% (active ingredient) solution of sodium hypochlorite and 10 μl of detergent (Tween 20). Occasionally swirl an ear during sterilization and rinse three times with sterile distilled water. As an alternative, ears collected from greenhouse grown plants can be simply sterilized with a 2 min rinse of 70% ethanol.
3. Grasp the holder in the ear base and transfer the ear to a large sterile plate or other sterile surface. With a fine scalpel remove the upper part of the kernels of an entire ear (remove a flap of pericarp).
4. With a blunt spatula pick up an embryo which lies at the basal edge of the endosperm of the immature caryopsis (**Fig. 1a**).

3.1.3. Inoculation and Co-culture

1. Isolated IE (1.5–2.0 mm) are collected for 15 min in an *Agrobacterium* cell suspension in 1.5-ml microcentrifuge tubes. After 15 min of embryo isolation the microfuge tube is set aside for 5 min.
2. Remove the *Agrobacterium* suspension using a pipette with fine tip. Transfer the embryos to standard ½ MS co-culture medium. Flip the embryos so the scutellum is facing up (*see* **Note 7, Fig. 1b**). Keep the coculture plates in a growth chamber set at 24°C and dark for approximately 24 h. Transient expression of GFP in IE after co-culture with *Agrobacterium* can be seen on **Fig. 1c, d**.

3.1.4. Selection and Regeneration

The described selection protocol is recommended for selection of transformants resistant to glyphosate. The construct is pMON65375. (All the media are solidified with 3-g/l Phytagel).

1. *First selection.* After co-culture, transfer the embryos onto the first selection medium, MSW57/500 mg/l carbenicillin (C500)/0.1 mM glyphosate (Gly). Each plate (90 × 25 mm) may contain up to 25 embryos. Keep the plates in a growth chamber (27°C, dark) for approximately 2 weeks.
2. *Second selection.* Transfer callus pieces onto the second selection medium, MSW57/C500/0.25 mM Gly. Briefly separate each callus piece, which could already be more than 5 mm in diameter, but keep all the tissue from one piece together as one unit (event). Keep the plates in plastic boxes in a growth chamber (27°C, dark) for approximately 2 weeks. Callus formation and GFP expression during different stages of selection are demonstrated on **Fig. 1e–k**.
3. *BA pulse.* Transfer the callus from the second selection medium onto MSBA/C250/0.1 mM Gly. Each plate may contain 5–8 events. Move the plates to a transparent plastic box in a lighted growth chamber (16-h light, typically 80–100 μE, 27°C). Leave the cultures on this medium for 5–7 days.
4. *Regeneration.* At the end of BA pulse, transfer the callus onto $MSGR^-$/C250/0.1 mM Gly medium in plates and keep the plates in a growth chamber at 16 h light, 80–100 μE, 27°C.
5. *Shoot growth.* In approximately 2 weeks on the regeneration medium some callus pieces may have regenerated green shoots with or without roots. Those shoots should be healthy looking and easily distinguished from some small shoots, which are no longer growing or have been bleached (dying). Transfer the healthy shoots into a Phytatrays™ or culture tubes with $MSGR^-$ containing 0.1 mM glyphosate and solidified with 3 g/l phytagel. During transfer, try to remove callus tissue attached to the root area of the shoots.
6. *Transplanting to soil.* When plants have reached the lid and developed one or more healthy roots (usually after 1–3 weeks), they are big enough to be moved to soil. Remove plants gently and rinse in room temperature water to remove culture medium from the roots. Transplant T_0 plants into Jiffy® pots or peat pots containing MetroMix-200. Put plants in a growth chamber at 27°C, 70% humidity and low light intensity for 1–2 week.

3.2. Transformation of Seedling-Derived Callus

Different stages of callus induction and transformation are presented in **Fig. 2**. In general, the transformation frequency with callus used as initial explants for transformation is lower than with IE but still can be above 10%.

3.2.1. Seed Sterilization and Germination

Fungicide-treated seeds are recommended to use as a starting material. Several media can be used for germination of seeds. In order to achieve efficient callus induction from H99 or other line use MSVS34 medium for seed germination.

1. Put the seeds into sterile Petri plates (1–2 layers of seeds). 3–4 plates with seeds can be placed into one plastic vacuum desiccator.
2. Connect vacuum desiccator hose to a vacuum in a fume hood, place 200 ml of undiluted sodium hypochlorite (6.15% active ingredient) in a 200-ml glass beaker, and place the beaker in the center of the desiccator. Place plates with seeds into the desiccator and slightly open the lid of each Petri plate. Add 2 ml of concentrated HCl to the beaker.
3. Put the cover on the desiccator, turn on the vacuum, and leave on long enough to pull a vacuum. Usually it takes 1 min. (The noise heard when the vacuum is first applied should disappear). Shut off the vacuum valve and leave the desiccator for 8–12 h.
4. When done, release the vacuum, open the desiccator, and close the plates containing the seeds. This is best done in an M2 biosafety hood to minimize chance of re-contaminating the seeds and being exposed to chlorine gas.
5. In a culture hood, place 9 or 12 sterilized seeds in Petri dish or Phytatray™ with MSVS34 germination medium (*see* **Note 8**).
6. Hold 2–3 Petri dishes together with a rubber band, and keep the dishes in a transparent plastic box. (The rubber band will keep the Petri dish lid on as the seeds germinate).
7. Germinate seeds in a lighted growth chamber (16 h light, 80–100 µE, and 27–28°C) for 7 days.

3.2.2. Callus Induction and Propagation

For callus induction, use well developed 7-days-old seedlings.

1. In the culture hood open the plates of germinated seeds, cut the seedlings from the seeds and put approximately eight cut seedlings into an empty Petri dish.
2. Hold the seedling with a pair of forceps. Using a scalpel blade cut off the upper part of the seedling approximately 5 mm above and below the apical meristem (nodal area). *See* **Fig. 2a.**
3. Next, make a longitudinal cut through the entire stem piece (**Fig. 2b**).
4. Place each piece with the wounded surface down on semi-solid callus induction MSW57 medium, 10–16 pieces per plate.
5. Incubate plates in a lighted growth chamber (16 h light, 80–100 µE, and 28°C).
6. After 2–4 weeks, examine the plates under microscope. Transfer embryogenic callus to fresh MSW57 medium (*see* **Note 9**).

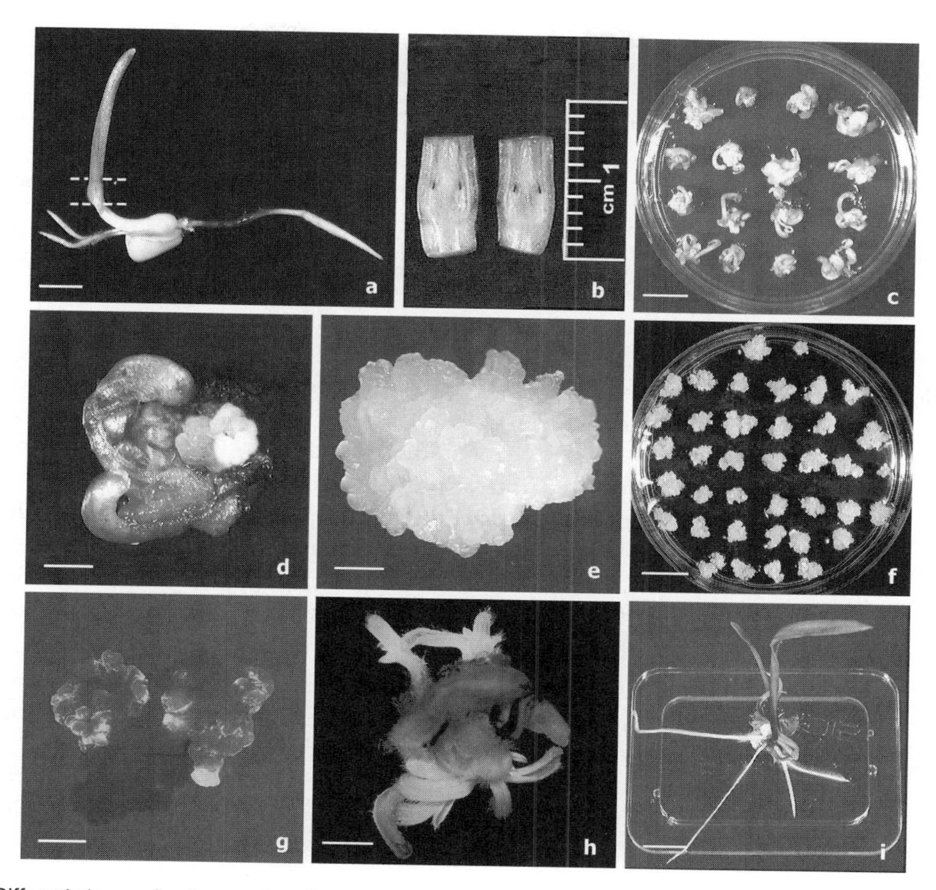

Fig. 2. Different stages of embryogenic callus induction of LH198 × Hill and its transformation using *Agrobacterium*. (**a**) Seven days-old corn seedling germinated on MSV34 medium (bar = 10 mm). (**b**) Nodal sections of seedling used as initial explants for callus induction. (**c**) Plate with nodal cuttings of seedling cultured for 4 weeks on MSW57 medium (bar = 16 mm). (**d**, **e**) Formation of embryogenic callus of on MSW57 medium and its morphology after establishment (bar = 4 mm and bar = 1.6 mm, respectively). (**f**) Petri plate with callus clumps suitable for *Agrobacterium*-mediated transformation (bar = 16 mm). (**g**) Transient GFP expression in seedling-derived callus; 2 days after inoculation (bar = 1.7 mm). (**h**) Regeneration of GFP positive shoots on MSGR⁻ medium (bar = 4 mm). (**i**) Rooting of plant in Phytatray™ with MSGR⁻ medium (bar = 16 mm) (*see Color Plates*).

Nodal cuttings with callus after 4 weeks of culture are shown on **Fig. 2c, d**.

7. Incubate *in darkness* at 28°C for 2–3 weeks.

3.2.3. Inoculation and Co-cultivation/Desiccation

Agrobacterium preparation is the same as for IE transformation but can be diluted to $O.D._{660}$ 0.5.

1. Collect in 50-ml sterile centrifuge tubes the callus pieces established and cultured on MSW57, which have been subcultured for approximately 1 week (*see* **Note 10**). Each tube may contain 10 ml of tissue. The callus used for transformation is shown on **Fig. 2e, f**.

2. Add to the tube enough prepared *Agrobacterium* solution to cover the tissue. Set the tube aside for 30 min.
3. Remove *Agrobacterium* suspension with a fine tip 5-ml pipette.
4. Dump the callus pieces onto two pieces of filter paper (Whatman #1, 8.5 cm in diameter) in a Petri dish (90 × 25 mm). Evenly disperse the callus pieces on the filter paper. Wrap the dishes with rubber bands and place them in a growth chamber (24°C, dark) for 2 days. (This is the coculture stage with a desiccation treatment.)

3.2.4. Selection and Regeneration

This protocol is recommended for selection with paromomycin or kanamycin. The callus, derived from nodal cuttings, during selection and regeneration is handled in the same manner as described in **Subheading 3.1.4**.

1. After the 2-day desiccation/coculture, transfer callus to MSW57 (500 mg/l carbenicillin and 100 mg/l paromomycin), about 15–25 callus pieces per plate. Culture for 3 weeks in the dark at 28°C. The transient expression of GFP in callus at this stage is shown on **Fig. 2 g**.
2. Transfer the callus pieces to MSW57 medium and break up each callus into 2–6 pieces. Culture for 3 weeks in the dark at 28°C. Resistant clones start to emerge at the end of this culture. If clones are not large enough for regeneration, leave on the same medium for up to 2 weeks without transferring. Alternately, an additional subculture on MSW57 can be done.
3. Transfer the resistant clones to MSBA (250 mg/l carbenicillin and 100 mg/l paromomycin) medium. Culture in the dark at 28°C for 7 days.
4. Transfer the tissues to MSGR$^-$ medium with 250 mg/l carbenicillin and 100 mg/l paromomycin medium in Phytatray™ or deep 90 × 25 mm Petri dishes. Place tissue at a density of 1–6 putative transformants/Phytatray™ or deep dish. Place in the high light growth chamber (16 h light, 80–100 μE, and 28°C) for approximately 4–6 weeks. Plants may need to be broken up for an additional subculture in 90 × 25 mm Petri dishes or Phytatray™. GFP expression in regenerating shoots and rooting of transgenic plant is shown in **Fig. 2 h, i**.
5. Plant the plantlets into soil (*see* **Note 11**).

4. Notes

1. Sigma, St. Louis, MO; Invitrogen, Carlsbad, CA; Phytotechnology Labs, Shawnee Mission, KS; Difco®: Becton Dickinson, Sparks, MD.

2. Antibiotics and acetosyringone stocks should be stored in a –20°C freezer in small aliquots. On average, they may be kept up to 6 months as a solution at –20°C or up to 1 week at 4°C.
3. LB plates should be stored upside-down inside a plastic bag at 4°C until ready to use. Do not store plates any longer than 1 month.
4. Keep glycerols on dry ice to prevent thawing.
5. After the initial 48 h growth period at 28°C streaked plates must be stored at 4°C, but not for more than 4 days. Plates older than this should not be used in transformation experiments due to potential plasmid rearrangements.
6. Avoid ears with any visible contamination or insect damage.
7. Orientation of IE during co-culture with *Agrobacterium* is critical. The embryos co-cultured with scutellum side up or scutellum side down have transient GFP expression. But only embryos co-cultured with scutellum side up have transient GFP expression on the upper surface of scutellum which gives embryogenic callus. Therefore, in most cases, only such embryos placed with scutellum side up give stable transformants.
8. Put fewer seeds per plate if the seed lot is highly contaminated.
9. It is critical to pick up only embryogenic callus. Do not take watery, friable callus which is not embryogenic. Identification of proper callus can be conducted under dissecting microscope.
10. Prolonged cultivation of embryogenic callus can reduce morphogenic potential of callus and can be a reason of somaclonal variation. Therefore for transformation it is recommended to use a callus which was initiated and maintained in culture no longer than 2.5–3 months.
11. Clean up the potting area and dispose of any transgenic material according to appropriate regulatory recommendations.

References

1. Hiei Y., Ohta S., Komari T., Kumashiro T. (1994) Efficient transformation of rice (*Oryza sativa* L.) mediated by *Agrobacterium* and sequence analysis of the boundaries of the T-DNA. *Plant. J.* 6, 271–282.
2. Ishida Y., Saito H., Ohta S., Hiei Y., Komari T., Kumashiro T. (1996) High efficiency transformation of maize (*Zea mays* L.) mediated by *Agrobacterium tumefaciens. Nat. Biotechnol.* 14, 745–750.
3. Zhao Z. Y., Gu W., Cai T., Pierce D. A. 2001. Nov. 9, (1999) Methods for *Agrobacterium-mediated transformation. United States Patent No. 5,981,840.*
4. Negrotto D., Jolley M., Beer S., Wenck A. R., Hansen G. (2000) The use of phosphomannose-isomerase as a selectable marker to recover transgenic maize plants (*Zea mays* L.) via *Agrobacterium* transformation. *Plant Cell Rep.* 19, 798–803.
5. Frame B.R., Shou H., Chikwamba R.K., Zhang Z., Xiang C., Fonger T.M., Pegg S.E.K., Li B., Nettleton D.S., Pei D., Wang K. (2002) Agrobacterium tumefaciens-mediated transformation of maize embryos using a standard binary vector system. *Plant Physiol.* 129, 13–22.
6. Cheng M., Fry J. E. (2000) An improved efficient *Agrobacterium*-mediated plant

transformation method. International Patent Publication WO 00/34491.

7. Cheng M., Hu T., Layton J., Liu C.-N., Fry J.E. (2003) Desiccation of plant tissues post- *Agrobacterium* infection enhances T-DNA delivery and increases stable transformation efficiency in wheat. *In Vitro Cell. Dev. Biol. Plant.* 39, 595–604.
8. Cheng M., Lowe B.A., Spencer T.M., Ye X., Armstrong C.L. (2004) Factors influencing *Agrobacterium*-mediated transformation of monocotyledonous species. *In Vitro Cell. Dev. Biol. Plant.* 40, 31–45.
9. Sidorov V., Gilbertson L., Addae P., Duncan D. (2006) *Agrobacterium*-mediated transformation of seedling-derived maize callus. *Plant Cell Rep.* 25, 320–328.
10. Zhang W., Subbarao S., Addae P., Shen A., Armstrong C., Peschke V., Gilbertson L. (2003) Cre/*lox* mediated marker gene excision in transgenic maize (*Zea mays* L.) plants. *Theor. Appl. Genet.* 107, 1157–1168.
11. Murashige T., Skoog F. (1962) A revised medium for rapid growth and bioassays with tobacco tissue cultures. *Physiol. Plant.* 15, 473–497.
12. Fromm M.E., Morrish F., Armstrong C., Williams R., Thomas J., Klein T.M. (1990) Inheritance and expression of chimeric genes in the progeny of transgenic maize plants. *Bio/Technology* 8, 833–839.

Chapter 5

Whiskers-Mediated Maize Transformation

Joseph F. Petolino and Nicole L. Arnold

Summary

There has been rapid progress in recent years in extending gene transfer capabilities to include plant species that fall outside the normal host range of *Agrobacterium*. Methods that allow direct DNA delivery into plant cells have contributed significantly to this expanded capability. Whiskers treatment is one means of delivering macromolecules, including DNA, to plant cells. Using relatively simple equipment and inexpensive materials, whiskers-mediated transformation of maize is possible. A critical prerequisite, however, is the establishment and maintenance of embryogenic tissue cultures as a source of totipotent, transformation-competent cells. Within hours of agitation in the presence of silicon carbide whiskers and DNA, embryogenic maize tissue cultures display transient gene expression, providing evidence for DNA uptake. Using appropriate selectable marker genes, following *in vitro* selection on inhibitory levels of a corresponding selection agent, stably transgenic tissue cultures can be generated from which fertile plants can be recovered. The timeline from whiskers treatment of embryogenic maize tissue cultures to fertile seed recovery is approximately 9 months, which is competitive with other methods of maize transformation.

Keywords: Transgenic production, Embryogenic tissue culture, Plant regeneration

1. Introduction

Silicon carbide whiskers are microfibers 10–80-μm long and 0.6-μm wide *(1)*. The vigorous agitation of plant cells in the presence of whiskers results in the formation of micron-sized openings in the cell wall, thereby allowing entry of macromolecules. Transient expression of reporter genes has been used to study conditions necessary for successful whiskers-mediated DNA uptake *(2–4)*. Selectable marker genes have been used to develop protocols for stable transgenic production *(5)*. Successful uptake and integration of DNA into maize cells following whiskers treatment has

M. Paul Scott (ed.), *Methods in Molecular Biology: Transgenic Maize, vol. 526*

DOI: 10.1007/978-1-59745-494-0_5

been reported for embryogenic suspension cultures *(6, 7)* as well as friable embryogenic callus *(8)*.

Whiskers transformation requires (a) a reliable method of producing and maintaining tissue cultures with particular morphological features, (b) conditions that allow efficient DNA delivery to the appropriate cells without inhibiting their capacity to contribute to the future of the culture, and (c) an effective means of isolating and recovering rare integration events *(9)*. Since whiskers deliver DNA most effectively to exterior cell layers, tissue cultures with surface cells competent for DNA uptake and integration as well as continued proliferation and *de novo* meristem formation are required. Embryogenic tissue cultures of maize meet these requirements *(10)*. In addition, temporary tissue plasmolysis using osmotic pretreatment is required as a means of mitigating the effects of whiskers-mediated cell wall damage *(11)*. The use of a functional selectable marker gene in combination with an effective selection agent is necessary for transgenic isolation and recovery *(12)*. The protocol described herein is based on the use of phophinothricin acetyltransferase (*pat*) as the selectable marker gene; however, alternate marker genes can be facilely utilized via empirical determination. Once established, plant regeneration from embryogenic maize tissue cultures is an exercise in media manipulation. High levels of auxin are used to initiate and maintain embryogenesis while auxin reduction in combination with cytokinin exposure is used for shoot and root formation *(13)*.

2. Materials

2.1. Culture Media

1. "15Ag10" medium (callus initiation): N6 salts and vitamins *(14)*, 1.0 mg/L 2,4-D, 20 g/L sucrose, 100 mg/L casein hydrolysate (enzymatic digest), 25 mM L-proline, 10 mg/L $AgNO_3$, and 2.5 g/L gellan gum powder (PhytoTechnology Laboratories, Shawnee Mission, KS) adjusted to pH 5.8.
2. "4" medium (callus maintenance): N6 salts and vitamins *(14)*, 1.0 mg/L 2,4-D, 20 g/L sucrose, 100 mg/L casein hydrolysate (enzymatic digest), 6 mM L-proline, and 2.5 g/L gellan gum powder adjusted to pH 5.8.
3. "H9CP+" medium (suspension culture): MS basal salts *(15)*, modified MS vitamins containing tenfold less nicotinic acid and fivefold higher thiamine–HCL, 2.0 mg/L 2,4-D, 2.0 mg/L α-naphthaleneacetic acid (NAA), 30 mg/L sucrose, 200 mg/L casein hydrolysate (acid digest), 100 mg/L myo-inositol, 6 mM L-proline adjusted to pH 6.0, with 5% (v/v) coconut water (C5915, Sigma Chemical Co., St. Louis, MO) added just before subculture.

4. "GN6" medium (whiskers treatment): N6 salts and 100 mg/L myo vitamins *(14)*, 2.0 mg/L 2,4-D and 30 g/L sucrose adjusted to pH 6.0.
5. "GN6S/M" medium (osmotic pretreatment): "GN6" medium with the addition of 45.5 g/L sorbitol, 45.5 g/L mannitol, adjusted to pH 6.0.
6. "GN6(1H)" medium (selection): N6 salts and vitamins *(14)*, 2.0 mg/L 2,4-D, 30 g/L sucrose, 100 mg/L myo-inositol, 1.0 mg/L bialaphos from Herbiace® (Meija Seika, Japan), and 2.5 g/L gellan gum powder adjusted to pH 6.0.
7. "GN6(1H)-agarose medium" (embedding): "GN6(1H)" medium with the substitution of 7 g/L SeaPlaque® agarose (BMA, Rockland, ME) adjusted to pH 6.0 autoclaved for only 10 min at 121°C.
8. "28(1H)" medium (shoot induction): MS salts and vitamins *(15)*, 30 g/L sucrose, 5 mg/L benzylaminopurine, 0.025 mg/L 2,4-D, 1.0 mg/L bialaphos from Herbiace®, and 2.5 g/L gellan gum powder adjusted to pH 5.7.
9. "36(1H)" medium (regeneration): MS salts and vitamins *(15)*, 30 g/L sucrose, 1 mg/L bialaphos from Herbiace®, and 2.5 g/L gellan gum powder adjusted to pH 5.7.
10. "SHGA" medium (rooting): SH salts and vitamins *(16)*, 1.0 g/L myo-inositol, 10 g/L sucrose, and 2.0 g/L gellan gum powder adjusted to pH 5.8.

2.2. Whiskers

1. In a chemical fume hood, 405 mg of dry silicon carbide whiskers (Silar SC-9, Advanced Composite Materials, Greer, SC) can be transferred to 30-mL polypropylene centrifuge tubes (Fisher Scientific, Pittsburgh, PA). Gloves and a respirator must be worn while weighing and transfer are performed, and damp paper towels should be spread out so as to immobilize any spilled whiskers, which, when dry, represent a serious respiratory hazard. The centrifuge tube containing whiskers can be autoclaved on a slow release cycle and stored at room temperature.
2. On the day of use, a 50 mg/mL suspension of whiskers can be prepared by adding 8.1 mL of "GN6S/M" medium to 405 mg of sterile whiskers and mixed for 60 s using a Vortex Genie 2® (Fisher Scientific, Pittsburgh, PA) set at maximum speed.

2.3. Supplies and Equipment

1. 250-mL centrifuge bottles (05-433A, Fisher Scientific).
2. Modified paint mixer (Red Devil Equipment Co., Minneapolis, MN) in which the paint can clamp assembly is retrofitted to hold a centrifuge bottle **Fig. 1**).
3. Glass filter holder (XX10-047-30, Fisher Scientific).
4. 500-mL filter flask (#10-180E, Fisher Scientific).

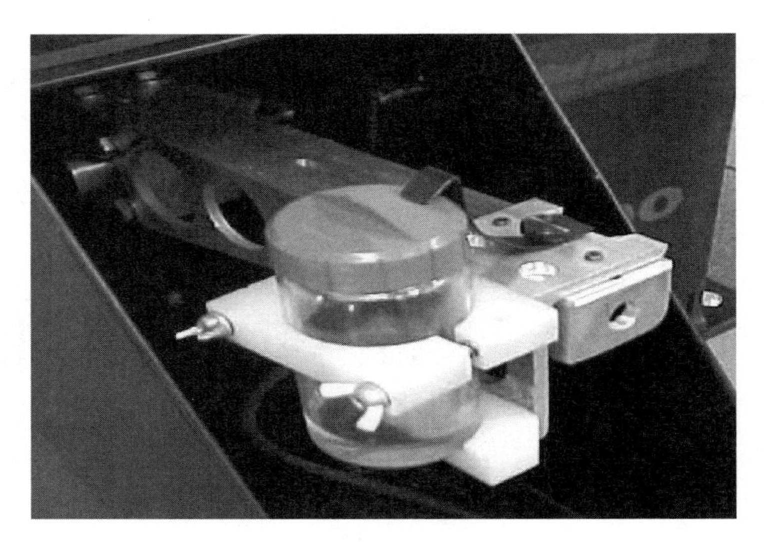

Fig. 1. Closeup of modified Red Devil paint mixer with centrifuge bottle attachment (*see Color Plates*).

5. Disposable 50-mL polypropylene centrifuge tube (#05-539-13, Fisher Scientific).
6. Nescofilm® (Karlan Research Products Corp., Santa Rosa, CA).

3. Methods

3.1. Establishment and Maintenance of Embryogenic Tissue Cultures

1. Immature embryos of the genotype "Hi-II" can be produced by intermating two S_3 lines ("Hi-II parent A" and "Hi-II parent B") derived from a B73 × A188 cross *(17)*. Controlled mating of maize plants begins with covering the primary ear shoot of the seed parent with a waterproof bag before silk emergence to avoid unwanted pollination. Ear shoots can be found in the axil of the sixth or seventh leaf from the top. When the first silks emerge from the covered ear shoots, the tip of the husks and silks can be cut with a sharp knife and recovered. The next day, a thick brush of silks will have emerged. Pollen can be collected from mature anthers beginning on the upper third of the tassel of the male parent over the course of several days. Fresh anthers begin to extrude early in the morning, so pollen can be collected over the course of several of hours by covering tassels with a brown paper bag secured firmly with a paper clip. To complete the pollination the ear shoot bag is removed and pollen is sprinkled over the newly emergent silk brush. The pollen collection bag can then

be wrapped around the pollinated ear shoot and stem and stapled to allow seed formation.

2. When embryos are 1.0–1.5 mm in size (approximately 9–10 days postpollination), ears can be harvested and surface sterilized by scrubbing with Liqui-Nox® soap (Alconox, Inc., White Plains, NY), immersed in 70% ethanol for 2–3 min, and then immersed in 20% commercial bleach (0.1% sodium hypochlorite) for 20–30 min.
3. Ears can be rinsed in sterile, distilled water, and immature zygotic embryos are aseptically excised and cultured in the dark at 28°C on 15Ag10 medium with the scutellum facing away from the medium (*see* **Note 1**).
4. After 2–3 weeks, friable callus with numerous globular and elongated somatic embryos appears on the scutellar surface of each plated embryo.
5. Embryogenic tissue can be selectively transferred at biweekly intervals using a dissecting microscope onto fresh "15Ag10" medium for approximately 6 additional weeks, and then transferred to "4" medium at biweekly intervals for approximately 8 additional weeks, continuing selection for embryogenic tissue morphology (*see* **Note 3**, **Fig. 2**).
6. Embryogenic suspension cultures can be initiated by transferring approximately 3 mL packed cell volume (PCV) of embryogenic callus tissue originating from a single immature embryo to approximately 30 mL of "H9CP+" medium (*see* **Note 3**). Developing suspension cultures can be maintained in the dark in 125-mL Erlenmeyer flasks in a temperature-controlled

Fig. 2. Friable embryogenic callus forming on the scutellar surface of immature embryos 2 weeks after culture on "15Ag10" medium (*see Color Plates*).

shaker (Innova® 4,330 Refrigerated Incubator, New Brunswick Scientific Co., Inc. Edison, NJ) set at 125 rpm at 28°C. Suspension cultures typically become established within 2–3 months after initiation into liquid medium. During establishment, suspensions can be subcultured every 3–4 days by adding 3-mL PCV of cells and 7 mL of "spent" (conditioned) medium to 20 mL of fresh "H9CP+" medium using a wide-bore pipette. Once the tissue starts consistently doubling in growth over this time period, suspension cultures can be scaled-up and maintained in 500-mL flasks whereby 12-mL PCV of cells and 28-mL conditioned medium is transferred into 80 mL of fresh "H9CP+" medium.

3.2. Whiskers-Mediated DNA Delivery

1. Approximately 24 h prior to transformation, 12-mL PCV and 28 mL of "spent" (conditioned) medium can be subcultured into 80 mL of fresh "GN6" medium in a 500-mL Erlenmeyer flask maintained at 125 rpm and 28°C. Repeating this procedure two more times results in 36-mL PCV distributed across three flasks.
2. After 24 h, the media in each flask can be removed and replaced with 72 mL of "GN6S/M" medium and incubated in the dark for 30–35 min at 125 rpm and 28°C after which the contents of the three flasks can be pooled and transferred to a 250-mL centrifuge bottle.
3. Once the cells have settled to the bottom of the centrifuge bottle, all but approximately 14 mL of medium can be drawn off and saved in a 1-L flask for later use. At this time, the 8.1 mL of 5% whiskers suspension can be added along with 170 μg of plasmid DNA.
4. Once the whiskers and DNA are added, the bottle is immediately placed onto a modified paint mixer and agitated for 10 s (*see* **Note 4**).
5. Immediately following agitation, the cocktail of cells, media, whiskers, and DNA can be added back to 1-L flask along with 125 mL of fresh "GN6" medium and allowed to recover for 2 h at 125 rpm and 28°C.

3.3. Transgenic Event Selection

1. Once the cells have been allowed to recover, 6-mL aliquots can be filtered onto 5.5-cm Whatman #4 filter paper using a glass filter holder and a 500-mL filter flask connected to a house vacuum line such that approximately 60 filters are obtained per bottle (**Fig. 3**).
2. Filters can be placed onto 60 mm × 20 mm Petri dishes containing "GN6" medium solidified with 2.5 g/L gellan gum powder and maintained at 28°C in the dark.
3. After 1 week, filter papers can be transferred to 60 mm × 20 mm Petri dishes containing "GN6(1H)" medium for an additional week in the dark at 28°C.

Fig. 3. Closeup of whiskers-treated tissue collected using a glass filter holder and filter flask connected to a house vacuum line (*see Color Plates*).

4. At this point (2 weeks postwhiskers treatment), the tissue can be embedded by scraping one-half of the cells from each filter into 3.0 mL of "GN6(1H)-agarose" medium held at 37–38°C in a disposable 50-mL polypropylene centrifuge tube. The tissue can be broken up with a spatula and the 3 mL of tissue/agarose medium is spread evenly onto the surface of 100 mm × 15 mm Petri dish containing "GN6(1H)" medium. This procedure can be repeated for the other half of the filter.
5. Once embedded, the plates can be sealed with Nescofilm® and cultured at 28°C in the dark for up to 10 weeks. Transgenic isolates can be removed from the embedded plates and transferred to fresh "GN6(1H)" medium in 60 mm × 20 mm Petri dishes. Sustained growth on the selection agent is an indication of stable transformation and warrants further analysis.

3.4. Plant Regeneration

1. Plants can be regenerated from transgenic cultures by first transferring them to "28" medium for 1 week in low light (14 $\mu E/m^2s$) with a 16-h photoperiod followed by 1 week in higher light (89 $\mu E/m^2s$) before being transferred to "36" medium for shoot formation.
2. Once shoots reach approximately 3–5 cm in length, they can be individually transferred into 150 mm × 25 mm glass culture tubes containing "SHGA" medium.
3. Once plantlets develop a sufficient root system and the shoots reach the top of the tube, they can be transplanted into 10-cm pots containing Metro-Mix 360® (Sun Gro Horticulture,

Bellevue, WA) and transferred to the greenhouse. The plantlets can be fully or partially covered with clear plastic cups for 2–7 days, and then transplanted at the 3–5 leaf stage into 5-gal pots containing a mixture of 95% Metro-Mix 360® and 5% clay/loam soil.

4. Plants can be grown to maturity in a greenhouse with supplemental lighting provided by metal halide and/or high-pressure sodium lamps. Healthy plant growth requires a 13–16-h photoperiod with day/night temperatures of about 30/20°C.
5. Self- and/or cross-pollinations can be performed as previously described to generate transgenic progeny.

4. Notes

1. Immature embryos will reach appropriate developmental stage (1.0–1.5 mm) in about 9–10 days; however, temperature and sunlight can have a large effect on timing. Immature embryo size can be checked by peeling back husks and sampling before ear harvest.
2. Prior to the establishment of suspension cultures, it is important at each subculture passage to select callus tissue of the correct morphology. Nonembryogenic callus is made up of soft, granular, translucent tissue, which is composed of elongated, vacuolated cells. Tissue of this type should be avoided. Tissue with a high density of finger-like projections covering the callus surface with a friable (crumbly) texture is the most amenable for initiating suspension cultures. Suspension culture age is critical. Cultures that are too young (<6 months) will not be as amenable to transformation and those that are too old (>12 months) will not be as regenerable and fertile.
3. Bacterial and fungal contamination can be unwelcome side effects of the *in vitro* process. As such, assessing the established suspensions by periodically plating aliquots onto a sterility test medium represents one way to monitor the cultures' health. In addition, allowing media to sit at room temperature for several days prior to use can also provide advance indication of contamination and thus potentially reduce the risk of experimental loss.
4. The whiskers treatment described here involves a rather large volume of suspension culture cells (36 PCV) agitated on a modified commercial-scale paint mixer. Smaller volumes of cells such as 2-mL PCV can be agitated in conical centrifuge tubes using a dental amalgamator such as a Vari-Mix II (Estrada Dental Co., Rancho Cucamonga, CA) or in Eppendorf tubes using a Vortex Genie 2®.

References

1. Coffee, R. and Dunwell, J.M. (1994) Transformation of plant cells. *United States Patent* 5302523, issued 12 April 1994.
2. Kaeppler, H.F, Gu, W., Somers, D.A., Rines, H.W., and Cockburn, H.F. (1990) Silicon carbide fiber-mediated DNA delivery into plant cells. *Plant Cell Rep.* 8:415–418.
3. Serik, O., Ainur, I., Murat, K., Tetsuo, M., and Masaki, I. (1996) Silicon carbide-mediated DNA delivery into cells of wheat (*Triticum aestivum* L.) mature embryos. *Plant Cell Rep.* 16:133–136.
4. Nagatani, N., Honda, H., Shimada, and T., Kobayashi. (1997) DNA delivery into rice cells and transformation of cell suspension cultures. *Biotechnol. Tech.* 11:471–473.
5. Dalton, S.J., Bettany, A.J.E., Timms, E., and Morris, P. (1998) Transgenic plants of *Lolium multiflorum, Lolium perenne, Festuca arundiacea, and Agrostis stolonifera* by silicon carbide fibre-mediated transformation of cell suspension cultures. *Plant Sci.* 132:31–43.
6. Frame, B.R., Drayton, P.R., Bagnall, S.V., Lewnau, C.J., Bullock, W.P., Wilson, H.M., Dunwell, J.M., Thompson, J.A., and Wang, K. (1994) Production of fertile transgenic maize plants by silicon carbide whisker-mediated transformation. *Plant J.* 6:941–948.
7. Thompson, J.A., Drayton, P.R., Frame, B.R., Wang, K., Dunwell, J.M. (1995) Maize transformation utilizing silicon carbide whiskers: a review. *Euphitica* 85:75–80.
8. Petolino, J.F., Hopkins, N.L., Kosegi, B.D., and Skokut, M. (2000) Whisker-mediated transformation of embryogenic callus of maize. *Plant Cell Rep.* 19:781–786.
9. Petolino, J.F. (2002) Direct DNA delivery into intact cells and tissues, in transgenic plants and crops (Khachatourians, G.C., McHughen, A., Scorza, R., Nip, W.K., and Hui, Y.H., eds.), Marcel-Dekker, New York, pp 137–143.
10. Welter, M.E., Clayton, D.S., Miller, M.A., and Petolino, J.F. (1995) Morphotypes of friable embryogenic maize callus. *Plant Cell Rep.* 14:725–729.
11. Vain, P., McMullen, M.D., and Finer, J.J. (1993) Osmotic treatment enhances particle bombardment-mediated transient and stable transformation of maize. *Plant Cell Rep.* 12:84–88.
12. DeBlock, M., Botterman, J., Vandewiele, M., Dockx, J., Thoen, C., Gossele, V., Rao, N., Thompson, C., Van Montagu, M., Leemans, J. (1987) Engineering herbicide resistance in plants by expression of a detoxifying enzyme. *EMBO J.* 6:2513–2518.
13. Vasil, I.K. (1994) Molecular improvement of cereals. *Plant Mol. Biol.* 25:925–937.
14. Chu, C.C., Wang, C.C., Sun, C.S., Hsu, C., Yin, K.C., Chu, C.Y., and Bi, F.Y. (1975) Establishment of an efficient medium for anther culture of rice through comparative experiments on the nitrogen sources. *Sci. Sinica.* 18:659–668.
15. Murashige, T. and Skoog, F. (1962) A revised medium for rapid growth and bioassays with tobacco tissue cultures. *Physiol. Plant.* 15:473–497.
16. Schenk, R.U. and Hildebrandt, A.C. (1972) Medium and techniques for induction and growth of monocotyledonous and dicotyledonous plant cell cultures. *Can J. Bot.* 50:199–204.
17. Armstrong, C.L., Green, C.E., and Phillips, R.L. (1991) Development and availability of germplasm with high Type II culture formation response. *Maize Coop. News Lett.* 65:92–93.

Part III

Transgenic Maize in Research

Chapter 6

Methods for Generation and Analysis of Fluorescent Protein-Tagged Maize Lines

Amitabh Mohanty, Yan Yang, Anding Luo, Anne W. Sylvester, and David Jackson

Summary

The use of fluorescent proteins to localize gene products in living cells has revolutionized cell biology. Although maize has excellent genetics resources, the use of fluorescent proteins in maize cell biology has not been well developed. To date, protein localization in this species has mostly been performed using immunolocalization with specific antibodies, when available, or by overexpression of fluorescent protein fusions. Localization of tagged proteins using native regulatory elements has the advantage that it is less likely to generate artifactual results, and also reports tissue-specific expression patterns for the gene of interest. Fluorescent protein tags can also be used for other applications, such as protein–protein interaction studies and purification of protein complexes. This chapter describes methods to generate and characterize fluorescent protein-tagged maize lines driven by their native regulatory elements.

Keywords: Confocal microscopy, Gateway cloning™, Green fluorescent protein (GFP), Yellow FP (YFP), Red FP (RFP), Maize, Native regulatory elements, Protein localization, Tissue-specific expression, Triple-template PCR (TTPCR)

1. Introduction

Over the past several years, tremendous progress has been made in obtaining genome sequences of many organisms. A first draft of the maize genome sequence for inbred line B73 was released in 2008, adding significantly to current genomic resources. As with all newly released genomes, however, understanding function of annotated genes remains a challenging task. Functional studies tend to lag behind the generation of sequence data, in part due to the need to develop new high-throughput tools for

M. Paul Scott (ed.), *Methods in Molecular Biology: Transgenic Maize, vol. 526*

DOI: 10.1007/978-1-59745-494-0_6

investigation. For example, although the *Arabidopsis* genome was completed in 2001, the functions of a large number of its genes remain unknown. Recent development of protein tagging methods will greatly facilitate functional studies by providing important localization information and by means of studying *in vivo* protein dynamics. Localization of gene products is now a priority for advancing postgenomic study in several organisms.

As in all eukaryotes, maize cells consist of discrete compartments, where proteins carry out a diverse array of specialized functions. The localization of proteins to specific compartments provides essential information about their biological activity. Certain proteins can also exhibit dynamic behavior that includes movement within and between compartments, e.g., from cytosol to nucleus, in response to specific signals. Traditionally, protein localization studies have been performed with antibodies generated against a protein of interest. However, generation of antibodies is a time-consuming process that is not always successful and is not easily amenable to high-throughput approaches. Immunolocalization also usually requires tissue fixation, causing potential localization artifacts and also preventing *in vivo* study of protein dynamics. The discovery and use of fluorescent proteins, such as the green fluorescent protein (GFP), has revolutionized the way protein localization is performed *(1)*. Fluorescent protein (FP) fusions allow analysis of dynamic localization patterns in real time. Over the last several years, a number of different color fluorescent proteins have been developed, including yellow FP (YFP), cyan FP (CFP), red FP (RFP), and others *(2, 3)*. Some of these proteins have improved spectral properties, allowing analysis of fusion proteins for a longer period of time and permitting their use in photobleaching experiments *(4)*. Others are less sensitive to pH, and other physiological parameters, making them more suitable for use in a variety of cellular contexts *(5)*. This ever-expanding tool kit can be used for simultaneous analysis of several proteins tagged with different color fluorescent proteins. Additionally, FP-tagged proteins can be used in protein–protein interaction studies by bioluminescence resonance energy transfer (BRET) or fluorescence resonance energy transfer (FRET) *(6)*.

High-throughput analyses of FP fusion proteins in *Arabidopsis* have been performed by overexpressing cDNA-GFP fusions driven by strong constitutive promoters *(7, 8)*. Although useful, this approach has inherent limitations, as it does not report tissue-specificity, and overexpression of multimeric proteins may disrupt the complex. Furthermore, overexpression can lead to protein aggregation and/or mislocalization. We therefore developed approaches to generate and analyze FP fusion proteins that are expressed by native regulatory elements *(9)*. This approach overcomes the problems associated with constitutive overexpression, and allows the analysis of fusion proteins in their native tissue or cell type and at native expression levels.

The application of fluorescent proteins in maize cell biology has been limited up to now, and has often involved transient overexpression in heterologous systems *(10–20)*. Recently, we have applied semihigh-throughput methods using Gateway recombination cloning *(21)* to fill this gap. Here, we describe methods for generation and analysis of maize FP-tagged lines driven by native regulatory elements in maize. Sample images from our work are available at our website (http://maize.tigr.org/cellgenomics/index.shtml).

2. Materials

2.1. Identification of Target Genes for Tagging and PCR Primer Design

1. PCR primers, as needed, can be obtained from any supplier.

2.2. Maize Genomic DNA Isolation

1. DNeasy® Plant Mini kit (QIAGEN).
2. Maize leaves (e.g., B73 inbred).
3. TE Buffer: 10 mM Tris–HCl, pH 8.0, 1 mM EDTA.

2.3. PCR Reactions for Amplification of Maize Genes for Tagging

1. KOD Hot Start DNA polymerase (Novagen).
2. Betaine (5M, Sigma).
3. DMSO (Sigma).
4. GFX gel extraction kit (GE Healthcare).
5. Primers for amplification of target genes by PCR (P1, P2, P3, and P4 primers; *see* **Subheading 3.1**).

2.4. Generation of Fluorescent Protein Tags

1. PCR primers and reagents, as described in **Subheading 2.3**, and methods.
2. Fluorescent protein clones, available from various companies (Clontech, Invitrogen, etc.) as well as from academic labs.

2.5. "Triple Template" (TT)-PCR and Generation of Gateway Entry Clones

1. KOD Hot Start DNA polymerase (Novagen).
2. DMSO (Sigma).
3. Betaine (5M, Sigma).
4. pDONR 207 (Invitrogen) or another gateway donor plasmid.
5. BP enzyme reaction mix (Invitrogen).
6. DH10B or DH5a electrocompetent *E. coli* cells.
7. LB plates containing gentamycin (7 mg/L).

8. LB medium
9. LB medium containing gentamycin (7 mg/L) or appropriate antibiotic.
10. Plasmid isolation kit (Promega).
11. Restriction enzymes as needed.

2.6. Generation of Gateway Binary Clones and Their Verification

1. Gateway LR clonase enzyme mix (Invitrogen).
2. Plasmids: pAM1006 (AM and DJ, unpublished), or another gateway binary vector.
3. Topoisomerase I (Invitrogen).
4. DH10B or DH5α electrocompetent *E. coli* cells.
5. LB plates containing spectinomycin (100 mg/L).
6. LB medium.
7. LB medium containing spectinomycin (100 mg/L).
8. KOD Hot Start DNA polymerase (Novagen).
9. DMSO (Sigma).
10. Betaine (5M, Sigma).
11. Plasmid isolation kit (Promega).
12. Restriction enzymes as needed.

2.7. Alternative Method: Maize Gene Tagging by Multisite Gateway

1. PCR primers for amplification of maize gene and FP fragments.
2. MultiSite Gateway® Three-Fragment Vector Construction Kit (Invitrogen).
3. Other general cloning and PCR reagents, as described in previous sections.

2.8. Transformation of Binary Clones into Agrobacterium tumefaciens and Maize

1. *Agrobacterium tumefaciens* strain EHA101.
2. Spectinomycin (100 mg/mL stock solution).
3. Kanamycin (50 mg/mL stock solution).
4. Chloramphenicol (25 mg/mL stock solution).
5. LB plates containing spectinomycin (100 mg/L), kanamycin (50 mg/L), chloramphenicol (25 mg/L).
6. LB medium.
7. Reagents for plasmid isolation by standard methods.

2.9. Materials for Microscopy

1. Microscope slides.
2. Coverslips.
3. Vaseline, vacuum grease, or Valap.
4. Counter stains: Propidium iodide (Sigma Chemical Co.), DAPI, FM 4-64, DiOC6 (Molecular Probes).

5. Dissecting tools including single-edge razor blades, fine scalpels, reverse-action microdissection scissors (Fine Science Tools, Inc.).
6. Dissecting microscope equipped with UV light excitation, optional.
7. Wide-field epifluorescence microscope, ideally equipped with infinity corrected, high numerical aperture objective lenses and filter sets with narrow bandpass emission filters. Selection of narrow bandpass filters is useful to reduce interfering autofluorescence of plant tissues.
8. Confocal microscope.
9. Image analysis software, such as ImageJ (NIH, shareware).

3. Methods

3.1. Identification of Target Genes for Tagging and PCR Primer Design

1. Usually a gene ideal for tagging has been identified through forward genetic analysis or by homology to an interesting gene from another model system. For generation of native expression constructs, full-length genomic sequence is required. To obtain these sequences, we routinely search several maize databases, including the Assembled *Zea mays* sequences [AZM, http://maize.tigr.org/ *(22)*], Maize Assembled Genomic Islands [MAGI, http://magi.plantgenomics.iastate.edu/blast/blast_r.html *(23)*] and the nearly completed first draft of the maize genome sequence (http://www.maizegenome.org/), as well as occasional reports in Genbank. For tagging of the full-length gene with an FP, the full-length gene sequence should be available, including all intron and exon sequences, ~3 kb of 5′ ("promoter") sequence and ~1 kb of 3′ sequence *(9)*.
2. A standard protocol is to insert the FP tag at a default position of ~10 amino acids upstream of the stop codon, following methods established for Arabidopsis *(9)*. The rationale is to avoid masking C-terminal targeting signals (such as endoplasmic reticulum (ER) retention or peroxisomal signals). In addition, by avoiding the N-terminus, disruption of N-terminal targeting sequences or transit peptides is avoided. However, choice of tag insertion is case-dependent, and it should be based on information on functional domains from database searches (e.g., SMART, http://smart.embl-heidelberg.de/; Target P, http://www.cbs.dtu.dk/services/TargetP/). If a homolog of the gene of interest has been successfully tagged in another organism, this information is also used to choose the optimal tag insertion site. A schematic of the FP tagging procedure, called "Triple Template PCR" (TTPCR) is depicted in **Fig. 1** *(9, 24)*.

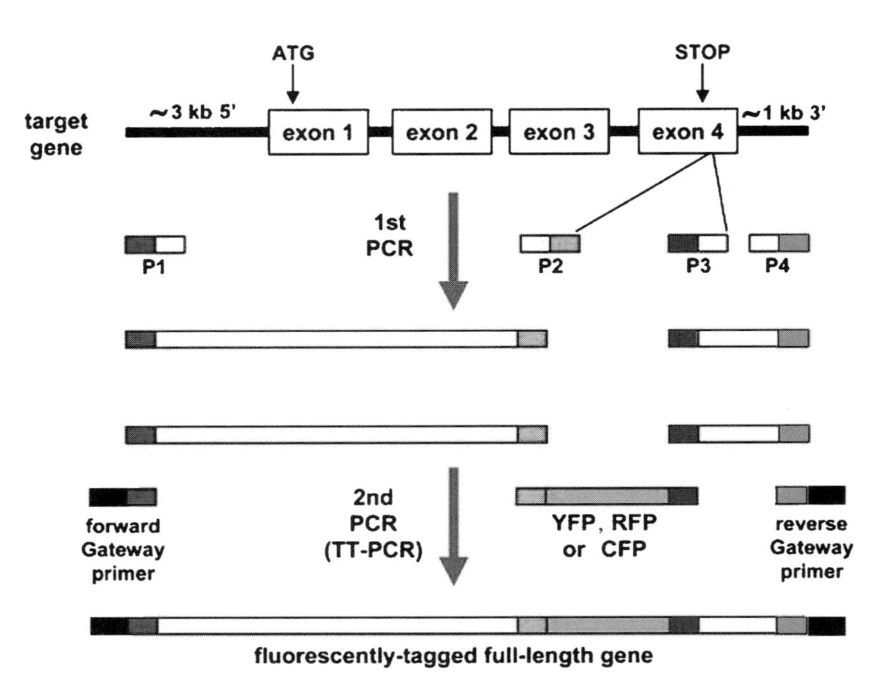

Fig. 1. Schematic of the TTPCR method that is used to tag genes with fluorescent proteins for expression using their native regulatory sequences. Primers are designated P1-P4 and forward or reverse gateway, as described in the methods. Figure is adapted from *(9)* (*see Color Plates*).

3. A set of four primers is designed for amplification of the target locus. Primers P1 and P2 amplify the 5′ regulatory regions and coding region, extending to the position where the FP tag will be inserted. The P3 and P4 primers are used to amplify the remainder of the gene from the tag insertion site and including the 3′ regulatory regions (**Fig.1**). Maize genomic DNA is used for amplification of P1-P2 and P3-P4 fragments. However, in cases where amplification from genomic DNA fails, maize BAC DNA clones, if available can be used as the PCR template (*see* **Note 1**). Primer design software PRIMER3 [http://frodo.wi.mit.edu/; *(25)]* is used for design of the P1-P4 primers. In general, the primer T_m should be 60–62°C. The P1 and P4 primers have linkers overlapping with the Gateway TTPCR primers in addition to the gene-specific sequences (sequences later) to allow cloning of the PCR product in gateway compatible vectors. Similarly the primers P2 and P3 contain gene-specific sequences as well as linkers that are complimentary to sequences from the YFP/CFP/RFP clones (*see* later) to allow incorporation of the FP tag into the TTPCR product (**Fig. 1**). The primer sequences are as follows:

P1 primer: GCTCGATCCACCTAGGCT + 18–25 gene-specific nucleotides.

P2 primer: CACAGCTCCACCTCCACCTCCAGGCCG-GCC + 18–25 gene-specific nucleotides.

P3 primer: TGCTGGTGCTGCTGCGGCCGCTGGGCC + 18–25 gene-specific nucleotides.

P4 primer: CGTAGCGAGACCACAGGA + 18–25 gene-specific nucleotides.

It is essential to maintain the correct open reading frame in the P2 and P3 primers, so that the plant protein sequence is maintained in frame with the FP tag. Example of primer sequences can be found at http://maize.tigr.org/cellgenomics/index.shtml.

3.2. Isolation of Maize Genomic DNA for PCR

1. The DNeasy® Plant Mini genomic DNA isolation kit (QIAGEN) is used for maize genomic DNA isolation, following manufacturer's instructions. Any method that produces high molecular weight genomic DNA is appropriate.
2. Genomic DNA is eluted with TE buffer and used directly for subsequent PCR reactions.

3.3. PCR Amplification of the Maize Gene Fragments for TTPCR Tagging

1. KOD Hot Start DNA polymerase (Novagen), a "proofreading" enzyme, is used for amplification of the maize genomic fragments. A typical PCR reaction is set up as follows:

- Maize genomic DNA (50–100 ng) = 1 μL
- 10× KOD enzyme buffer = 2.0 μL
- 10× dNTPs (2 mM each) = 2.0 μL
- MgSO4 (25 mM) = 1.2 μL
- Betaine (5M) = 4 μL
- DMSO = 0.4 μL
- Forward primer (10 μM) = 2.0 μL
- Reverse primer (10 μM) = 2.0 μL
- KOD DNA polymerase = 1 μL
- H2O = to bring up to 20 μL

Fixed annealing temperature PCR, or a "touchdown PCR" can be used as follows:

(a) Conditions for fixed annealing temperature PCR are as follows:

1. 94°C 2 min 30 s
2. 94°C 30 s
3. 58°C 30 s
4. 68°C X min (1 min per kb product length)
5. Go to step 2 and repeat for 30–32 cycles.
6. 68°C 10 min
7. 4°C forever

(b) Conditions for the touchdown PCR are as follows:

1. 94°C 2 min 30 s
2. 94°C 30 s

3. 63°C 30 s
4. (−1° per cycle)
5. 68°C X min (1 min per kb product length)
6. Go to step 2 and repeat six times
7. 94°C 30 s
8. 54°C 30 s
9. 68°C X min (1 min per kb product length)
10. Go to step 7 and repeat 23–25 times.
11. 68°C 10 min
12. 4°C forever

To get sufficient product for the subsequent TTPCR reactions, 3–4 PCR reactions (20 μL each) should be set up for the P1-P2 fragment and two PCR reactions for the P3-P4 fragment.

2. Check a small aliquot (1 μL/20) of the PCR product on an agarose gel (0.7% agarose in TBE buffer) to verify the correct size. If the size is correct, the remaining products are run on the gel and after separation the bands are carefully excised (*see* **Note 2**).
3. For gel extraction of the PCR product, the GFX gel extraction kit (GE healthcare) is used, following the manufacturer's instructions but with an additional drying step at 37°C for 5 min before elution of the DNA from the column.
4. The DNA is eluted in 25 μL of extraction buffer, which should give a fragment concentration of ~25–50 ng/μL.

3.4. Generation of Fluorescent Protein Tags for TTPCR

1. For gene tagging we routinely use the citrine variant of the Yellow Fluorescent Protein (YFP) *(5)*. Citrine-YFP has higher photostability than GFP and is less sensitive to pH and anions such as halides, thereby making it more suitable for use in a wide variety of subcellular contexts. We have also successfully used other fluorescent proteins such as cyan fluorescent protein (CFP; ECFP, Clontech) and the modified red fluorescent protein (mRFP1) *(26)*. We modified the fluorescent protein tags to remove start and stop codons and add flexible linker peptides flanking the ends allowing them to be used as either C- or N-terminal fusions or as internal fusions. These flexible linkers help to minimize folding interference between the target protein and the fluorescent protein. In addition, the linker peptide sequences contain an *Fse*I site at the 5′ end and an *Sfi*I site at the 3′ end. These restriction enzyme sites can be used in future to replace one fluorescent protein tag with another, or for addition of others, such as affinity purification tags for proteomics. We have generated CitrineYFP, ECFP, and mRFP1 clones with these linkers.

2. The fluorescent protein tag fragments are PCR amplified from the above plasmids using the following primers:

   Forward linker/FseI primer for YFP/CFP

   GGC CGG CCT GGA GGT GGA GGT GGA GCT GTG AGC

   Reverse linker/SfiI primer for YFP/CFP

   GGC CCC AGC GGC CGC AGC AGC ACC AGC AGG ATC

   Forward linker/FseI primer for RFP

   GGC CGG CCT GGA GGT GGA GGT GGA GCT GCC TCC TCC GAG GAC GTC ATC

   Reverse linker/SfiI primer for RFP

   GGC CCC AGC GGC CGC AGC AGC ACC AGC AGG ATC GGC GCC GGT GGA GTG GCG GCC

These primers are used to amplify the FP tag fragments, as follows:

Template plasmid = ~50 ng
1× KOD buffer = 2 μL
10× dNTPs (2 mM each) = 2 μL
MgSO4 (25 mM) = 1.2 μL
Forward primer (10 μM each) = 2 μL
Reverse primer (10 μM each) = 2 μL
KOD DNA polymerase = 1.0 μL
Total volume = 20 μL

PCR conditions:

94°C 2 min 30 s
94°C 30 s
56°C 30 s
72°C 1 min
Go to step 2, and repeat it 24 times.
72°C 10 min.
4°C forever.

3. The PCR products are run on a 1% agarose gel in TBE buffer, and the DNA fragment extracted is from the gel using the GFX gel extraction kit as described earlier (*see* **Note 2**). These fragments are called citrineYFP-TT, CFP-TT, and RFP-TT.

3.5. TT-PCR for Generation of the Maize Gene-FP Fusion, and Gateway Cloning

1. For the TT-PCR reaction, the following quantities should be used: ~100 ng of the P1-P2 PCR product, ~50 ng of the P3-P4 PCR product, and ~25 ng of the citrineYFP-TT, CFP-TT, or RFP-TT fragment. This PCR reaction leads to fusion of the FP tag with the two maize gene fragments to produce a full length tagged maize gene flanked by attB1 and attB2 recombination sites for cloning into gateway donor vectors *(9)*. The PCR conditions are the same as for generation of P1-P2 and P3-P4 fragments, as described earlier, remembering to add up all of the fragment lengths to calculate the extension time. In some cases it is necessary to use “gradient PCR” with a range

of annealing temperatures to optimize the PCR conditions for successful TTPCR.

The PCR primers for the TTPCR reaction are universal primers that contain gateway attB1 and attB2 sequences (underlined) and sequences corresponding to the nongene-specific parts of the P1 and P4 primers. The nucleotide sequences are given here:

Forward TTPCR gateway primer: GGGGACAAGTTTGTACAAAAAAGCAGGCTGCTCGATCCACCTAGGCT

Reverse TTPCR gateway primer: GGGGACCACTTTGTACAAGAAAGCTGGGTCGTAGCGAGACCACAGGA

Typically, 2–4 PCR reactions (25 μL each) are required to obtain enough PCR product for the subsequent BP cloning reaction.

A typical TTPCR reaction is as follows:

- P1-P2 fragment (~100 ng) = *x* μL
- P3-P4 fragment (~50 ng) = *y* μL
- YFP or CFP or RFP-TT fragment (~25 ng) = z μL
- 10× KOD enzyme buffer = 2.5 μL
- MgSO4 (25 mM) = 1.5 μL
- 10× dNTPs (2 mM each) = 2.5 μL
- Betaine (5M) = 5 μL
- DMSO = 0.5 μL
- Forward TTPCR gateway primer (10 μM) = 2.5 μL
- Reverse TTPCR gateway primer (10 μM) = 2.5 μL
- KOD enzyme = 1–1.5 μL
- H2O to bring up to 25 μL

2. A small aliquot of the reaction is run on an agarose gel (~0.8% in TBE) to verify that the correct size PCR product has been generated. The remainder of the product is then run on an agarose gel and the TTPCR fragment is extracted from the gel as described earlier (*see* **Note 2**).

3. The TTPCR fragment is cloned into a pDONR plasmid, such as pDONR207 (*see* **Note 3**), using the gateway BP recombination reaction (Invitrogen). The BP reaction promotes site-specific recombination between the attB sites of the TTPCR product and the attP sites of the vector, resulting in the production of a cloned fragment flanked by attL sites.

The BP reaction is set up as follows and incubated at 25°C overnight:

- TTPCR product = ~ 5 μL (150–250 ng)
- pDONR207 = 1 μL (~100 ng)
- BP reaction buffer = 2 μL
- BP enzyme (*see* **Note 4**) = 2 μL
- TE buffer (pH 8.0) to bring up to 10 μL

4. The reaction is stopped by addition of 1 μL of proteinase K followed by incubation at 37°C for 10 min.

5. Use a small aliquot (1–2 μL) of the reaction for transformation of electrocompetent DH10B *E. coli* cells. The rest of the reaction

mix can be stored at –20°C for later use, if required. After transformation by electroporation, add 800-μL LB medium, grow cells with shaking for ~1 h at 37°C, and then plate on LB plates containing gentamycin (7 mg/L) and incubate at 37°C overnight.

6. 4–8 colonies (*see* **Note 5**) from the transformation plate should be picked for PCR confirmation. Each colony is resuspended in 20 μL water, and 1 μL used as template for a PCR reaction using standard PCR conditions (*see* earlier sections). The remaining bacterial suspension is used to set up an overnight culture with the appropriate antibiotics. For each clone, set up a positive PCR control with diluted TTPCR product (~1 ng) as a template, and use empty vector DNA (e.g., pDONR207) as negative control. The P1 and P4 primers are used to test for presence of the full-length TTPCR product in the colony PCR reactions.
7. Once the correct clones have been identified, isolate the corresponding plasmid, for example, using a plasmid isolation kit (Promega). This "entry" plasmid should be further verified by restriction enzyme digestion and coding regions of the gene fusion, including the FP and linker regions, should be sequenced to control against mutations that may have been introduced during the PCR. We routinely sequence the FP linker regions first, since these are the most likely to contain mutations from the long primer synthesis. The cloning efficiency for the BP reaction is generally more than ~70%, but is lower for inserts larger than ~6 kb.

3.6. Transfer of Tagged Maize Genes to Binary Vectors by Gateway LR Reaction

1. Sequence-verified TTPCR entry clones are used for the LR reaction (Invitrogen). The LR reaction promotes a recombination reaction between the attL sites flanking the FP-tagged gene in the entry clone and attR sites in a gateway destination binary vector. We have generated a destination vector version of a common maize transformation vector, pTF101.1 *(27)*, and it is available on request (pAM1006, Mohanty A and Jackson D, unpublished).
   (a) A typical LR reaction is set up as follows and incubated at 25°C overnight:
      - Donor TTPCR clone ~250 ng
      - Binary vector (eg pAM1006) ~150 ng
      - LR reaction buffer = 2 μL
      - Topoisomerase I (7 U/μL) = 0.7 μL
      - LR enzyme (*see* **Note 4**) = 2 μL
      - TE buffer (pH 8.0) to bring up to 10 μL
2. Stop the reaction by addition of 1 μL of proteinase K followed by incubation at 37°c for 10 min.
3. Use a small aliquot (1–2 μL) of this reaction mix for transformation of electrocompetent DH10B cells. The rest of the reaction

mix can be stored at –20°C for later use, if needed. After transformation by electroporation, add 800-μL LB medium and grow cells for ~1 h at 37°C and then plate on spectinomycin (100 mg/L) plates followed by incubation at 37°C overnight.

4. Pick four colonies for PCR confirmation by colony PCR, as described earlier. Plasmids corresponding to the PCR positive clones can be prepared and digested with restriction enzymes for further verification. Cloning efficiency of the LR reaction is more than 85% for most of the maize clones we have generated.

3.7. Alternative Method: Fluorescent Protein Tagging Using Multisite Gateway® Three-Fragment Vector Construction Kit (Invitrogen)

As an alternative to gene tagging by TTPCR, we have also recently used the "Multisite Gateway Three-Fragment Vector Construction Kit" (Invitrogen). In this method, three DNA fragments are first cloned into different donor vectors in separate BP reactions. In the following LR reaction, site-specific recombination occurs among three different entry clones and the destination vector, to allow formation of the gene fusion, and its incorporation into the destination vector, in a single step. This method is particularly useful for construction of large fusion genes (>8 kb), where the TTPCR efficiency is usually low. This method requires two BP reactions for creation of one construct; however, since only one round of PCR reactions are needed, errors are reduced. In addition, once the entry clone of a certain fluorescent protein has been made, it can be used in construction of any fusion proteins to be fused with the same fluorescent tag. Since this method is new we do not describe it in detail here, but details can be found at the Invitrogen web site, and from DJ on request.

3.8. Transfer of Binary Clones to Agrobacterium and Maize

1. Binary plasmids are transferred to *Agrobacterium* (e.g., EHA101 strain) by electroporation. We use ~100 ng of plasmid DNA for ~50 μL of electrocompetent cells. After electroporation, add 800 μL of LB medium to the tubes and incubate at 28°C for 2 h with shaking.

2. Plate aliquots of 50 and 200 μL on LB plates containing spectinomycin (100 mg/L), kanamycin (50 mg/L), and chloramphenicol (25 mg/L) and incubate for 2–3 days at 28°C. Spectinomycin is used for selecting the binary plasmid, whereas the other two antibiotics are for the selection of the EHA101 *Agrobacterium* strain.

3. Pick single colonies and grow them for 2-3 days in 6-mL LB medium supplemented with above antibiotics with shaking at 28°C.

4. To verify the clones, the plasmids can be isolated from these cultures by a modified alkaline lysis method (*see* **Note 6**) and checked by restriction enzyme digestion.

5. Following clone verification, the constructs are transformed into maize to generate stable lines. Transgenic maize plants expressing

the FP tagged genes can be generated using public facilities, such as the Plant Transformation Facility (PTF) at Iowa State University. Constructs are mailed to the facility following requested protocols at http://www.agron.iastate.edu/ptf/index.aspx (*see* **step 9** in **Subheading 3.8** on Federal guidelines and procedures for working with transgenic maize).

6. Maize transformants are provided as seedlings on sterile Petri plates, regenerated from callus tissue from HiII lines (classified here as T_0 generation). Upon receipt, transfer the plates to growth chambers maintained at 25–28°C (16-h light period) until the roots and shoots are several cm long. Once acclimated in the growth chamber, the T_0 seedlings may be screened for expression (*see* **Subheading 3.9**).
7. Transfer seedlings to soil in small pots and cover with a plastic dome to maintain humidity for 3–4 days and encourage optimal root growth.
8. The established seedlings should be transferred to larger pots for growth and pollination in the greenhouse. To maintain adequate growth, greenhouse conditions should be optimized for maize. Tester lines (including inbreds such as B73 or mutants for complementation tests) to be crossed to the transgenic lines are planted every 3–4 days starting 2 weeks before the seedlings ship from the PTF and up to a week or 2 after they arrive. To confirm biological function of any tagged protein, its ability to complement the corresponding mutant phenotype is usually required.
9. Notes for working with transgenic maize: It is essential that all users of transgenic maize in the USA follow regulations found at http://www.aphis.usda.gov/. Arrangements should be made prior to shipping or working with transgenic maize. The online application procedure is available at http://www.aphis.usda.gov/permits/brs_epermits.shtml. Notifications take one to several months to approve, so should be applied for well in advance.

3.9. Screening of Transgenic lines, Analysis of Subcellular Localization, and In Vivo Dynamics

1. Primary screening: T_0 seedlings from the PTF can be screened prior to transplanting. A fluorescence dissecting microscope can be used to directly observe root tips on inverted plates for fluorescence. More detailed screening using a standard wide-field epifluorescence microscope (WM) will more definitively confirm positive events. For this primary screening, root tips should be collected under sterile conditions from T_0 seedlings, transferred to a drop of water, covered with a coverslip, and viewed with an epifluorescence WM equipped with appropriate filter cubes (*see* **step 2** in **Subheading 3.9** later). For proteins not expressed in root tips, leaf samples can be removed from the seedlings and similarly screened. Preliminary screening at the T_0 stage allows for selective planting of several positive

events to carry to T1 seed. However, for proteins expressed at later stages of development, all seedlings should be transplanted for secondary screening at later stages.

2. Secondary screening using standard wide-field epifluorescence microscopy (WM) is an excellent next choice for screening plants for tissue and developmentally specific expression of transgenes. WM is particularly useful for screening samples that can be hand sectioned, as detailed in **step 4** of **Subheading 3.9**. Screening can be accomplished using any high numerical aperture objective, and appropriate filter cubes for the FP tag. Natural plant cell autofluorescence may overlap with FP emission. Autofluorescence can be avoided by either using narrow bandpass filters or by merging images generated from several filter sets to detect overlap. **Figure 2** depicts the use of WM to detect localization patterns in maize leaves and roots using a Nikon E600 upright epifluorescence microscope equipped with cubes to detect cell wall autofluorescence (UV-2E/C, excitation at 340–380 nm, emission at 435–485 nm), YFP fluorescence (YFP HYQ, excitation at 490–510 nm, emission at 520–550 nm), and RFP or chlorophyll autofluorescence (G2E/C, excitation at 528–553 nm, emission at 590–650 nm).
3. Protein localization by confocal laser scanning microscopy (CLSM) is a method to optically section plant tissues, and is useful for working with thicker tissues such as leaf primordia, root and shoot apices (**Fig. 3**) or to localize FP tags with improved cellular resolution compared to WM. Tissues for confocal microscopy can be prepared as described earlier or by dissection of developing leaf tissue from within the whorl, or dissection of emerging ear or tassel inflorescences. Most institutions have a facility equipped with a CLSM, and training on the specific equipment should be obtained.
4. Specimen preparation and counter staining:
   (a) Root Tips: Remove a small clump of roots from established plants by hand, rinsing them free of soil with water. Identify small emergent lateral roots (2–3 mm in length). Avoid damaged root tips with brown ends. Snip the emergent tips with reverse action scissors and transfer to a drop of water on a glass slide. If the tip is greater than 2 mm in diameter, section longitudinally using a sharp razor blade. Rapid strokes ensure procuring a thin longitudinal section. Coverslips should be sealed with vaseline or vacuum grease on at least two sides to eliminate drying and to stabilize the tissue.

   (b) Hand cross sections of mature leaves: Most visible leaf tissue is too thick for microscopy without hand sectioning.

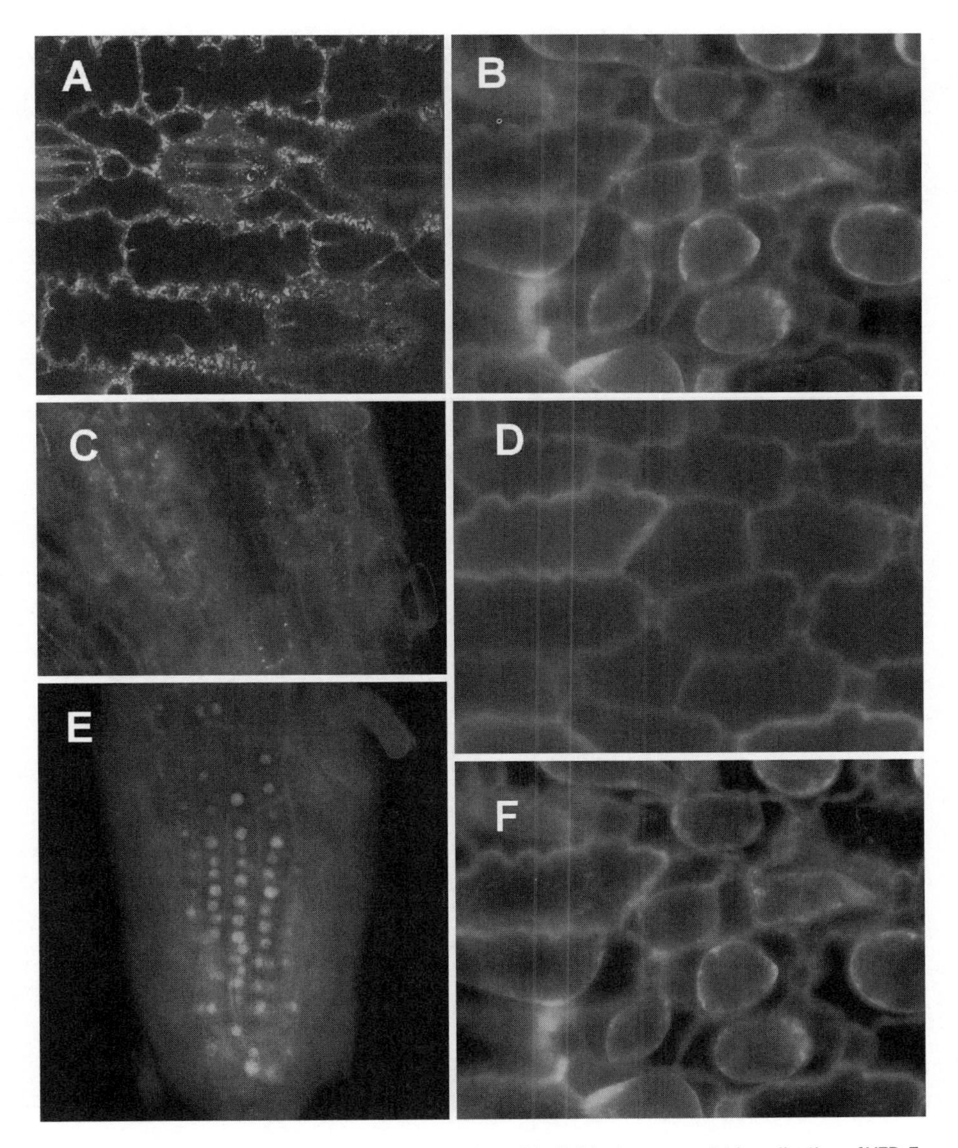

Fig. 2. Expression of fluorescent protein fusions in maize using wide-field microscopy. (a) Localization of YFP-ExpansinA1 in cytoplasm and wall space of hand paradermal sections of maize leaves, counterstained with PI. (**b**, **d**, **f**) Merging of images identify wall and cytoplasmic localization in plasmolyzed cells of YFP-ExpansinA1 in juvenile leaf cells [(**b**) YFP channel, (**d**) PI channel, (**f**) merged YFP and PI channels, AS and AL, unpublished]. (**c**) Localization of YFP tagged HSP22 to mitochondria in expanding root cells, merged with UV light-induced cell wall autofluorescence. Note punctate green fluorescence typical of mitochondrial distribution (AS and AL, unpublished). (**e**) Localization of YFP fusion of a DNA repair protein to nuclei of root tip cells, highlighted by merging with wall autofluorescence (AS, AL, and Cliff Weil, unpublished) (*see Color Plates*).

Hand sections can be readily obtained from mature tissues by rapidly sectioning with a fresh razor blade on a microscope slide. Cross sections will curl slightly at the end.

(c) Paradermal leaf sections: Single layers of epidermal tissue are particularly useful for imaging due to the absence of

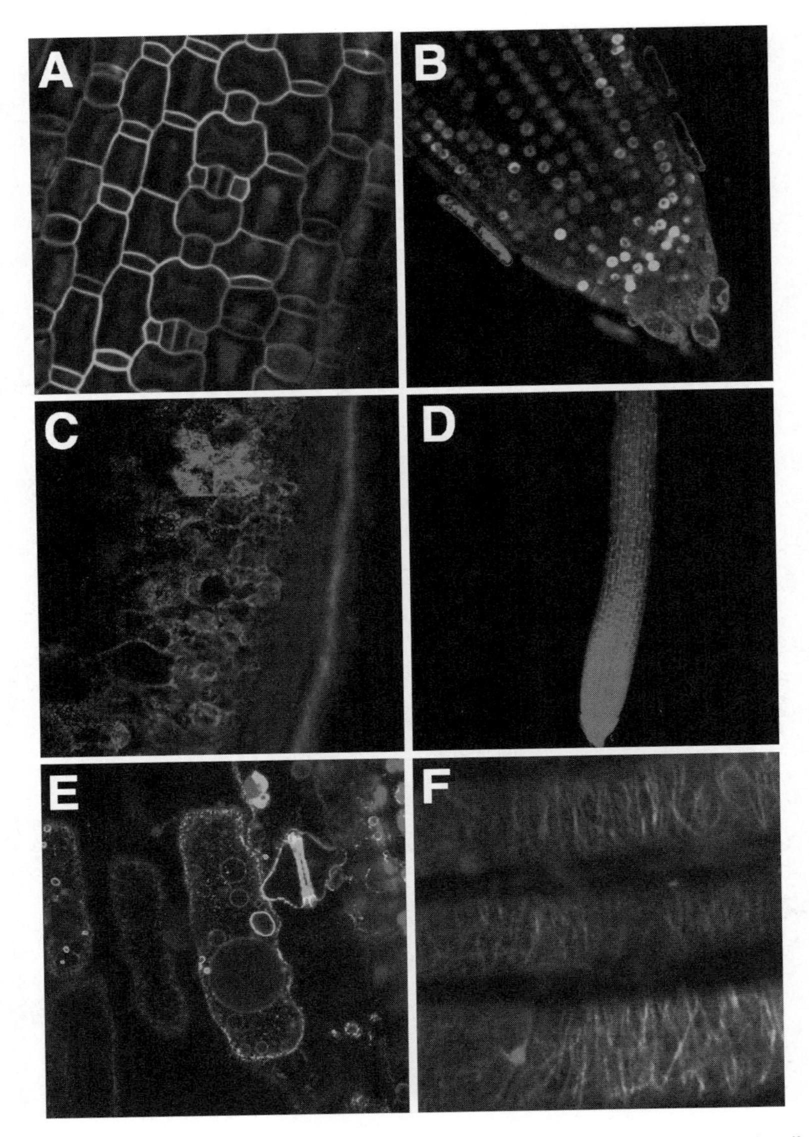

Fig. 3. Visualization of fluorescent protein fusions in maize using CLSM. (**a**) Plasma-membrane localization of a PIN-FORMED1-YFP fusion in maize leaf (YY, Andrea Gallavotti and DJ, unpublished). (**b**) Nuclear localization of a Histone H1-YFP fusion (in *green*) in root tip, cells are counterstained using FM4-64 dye (molecular probes) (AM and DJ, unpublished). (**c**) Expression of a zein-RFP fusion in developing endosperm tissues, note the punctate localization corresponding to protein bodies. Autofluorescence of the aleurone layer is in *green* (YY and DJ, unpublished). (**d**) Expression of a maize histidine phosphotransfer protein-RFP fusion in the root (AM and DJ, unpublished). (**e**) Localization of a tonoplast intrinsic protein-YFP fusion in membranes, presumably vacuolar, of leaf cells (AM and DJ, unpublished). (**f**) Expression of a tubulin-YFP fusion labels microtubules in maize root cells. Plasma membranes are counterstained using FM4-64 dye (molecular probes) (AM, AS and DJ, unpublished) (*see Color Plates*).

chlorophyll autofluorescence from most epidermal cells, except guard cells. An epidermal layer can be isolated by placing a 3 cm × 2 cm strip of leaf against a slide, and scraping gently with a razor blade, to remove one epidermal

Color Plates

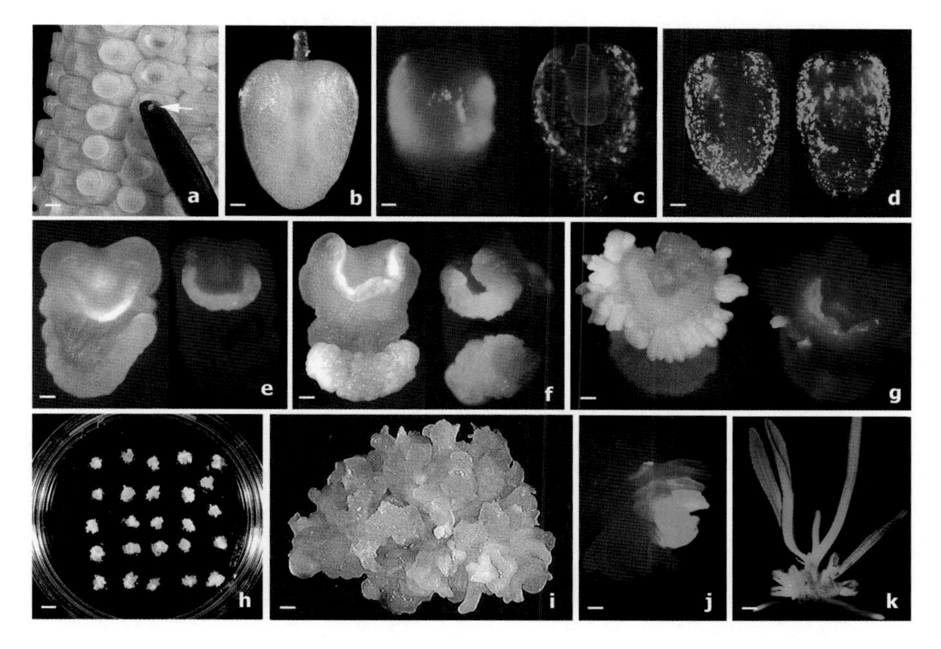

Chapter 4, Fig. 1. Different stages of *Agrobacterium*-mediated transformation of corn using freshly isolated immature embryos (IE). (**a**) Ear of corn with tops of kernels removed. *Arrow* indicates isolated IE at the end of spatula (bar = 3 mm). (**b**) Freshly isolated IE with scutellum surface up (bar = 0.16 mm). (**c**) Transient GFP expression in IE (scutellum up at *left*, and scutellum down at *right*; first day on selection medium), which were placed on medium during co-cultivation with scutellum side facing down. This is not correct orientation of IE during co-culture. These IE usually do not produce transgenic callus and plants. No GFP positive spots can be seen on surface of scutellum which forms embryogenic callus (bar = 0.18 mm). (**d**) Two IE at the first day on selection medium which had transient GFP expression on the upper surface of scutellum. During the co-culture with *Agrobacterium* the scutellum of embryos was facing up (bar = 0.2 mm). (**e**) IE 7 days on selection. IE were on selection medium with scutellum side up. To the *left* is IE in incandescent light, and at the *right* is IE at the *blue* light. Transient GFP expression is still visible at the area of callus formation (bar = 0.3 mm). (**f**) The same as on (**e**). Callus is forming at both side of IE (bar = 0.4 mm). (**g**) IE after 12 days on selection medium. To the *left* is IE in the incandescent light, and at the *right* is IE at the *blue* light. Callus area with stable GFP expression is visible (bar = 0.5 mm). (**h**) Individual calli derived from IE after 1 month of selection (bar = 6 mm). (**i**) Higher magnification of single callus line from plate shown on (**h**) (bar = 0.4 mm). (**j**) Stable expression of GFP in callus observed after 1 month of selection (bar = 0.3 mm). (**k**) GFP expression in regenerated plants obtained 3 months after inoculation (bar = 2 mm).

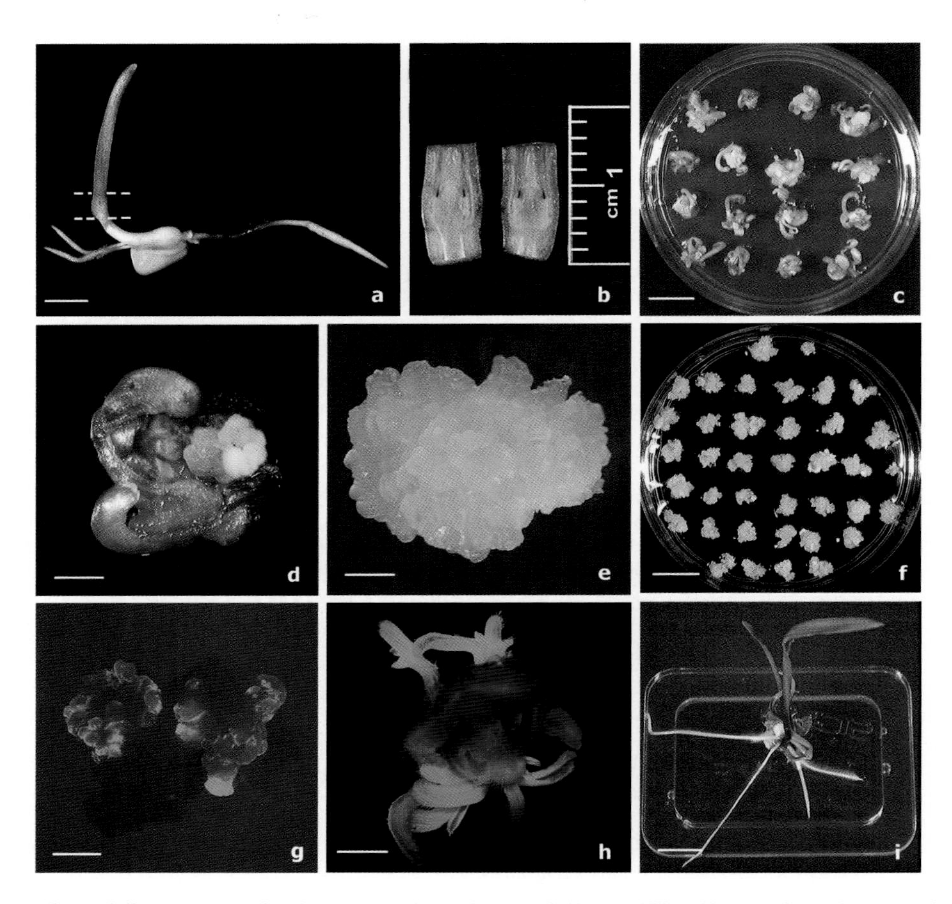

Chapter 4, Fig. 2. Different stages of embryogenic callus induction of LH198 × Hill and its transformation using *Agrobacterium*. (**a**) Seven days-old corn seedling germinated on MSV34 medium (bar = 10 mm). (**b**) Nodal sections of seedling used as initial explants for callus induction. (**c**) Plate with nodal cuttings of seedling cultured for 4 weeks on MSW57 medium (bar = 16 mm). (**d**, **e**) Formation of embryogenic callus of on MSW57 medium and its morphology after establishment (bar = 4 mm and bar = 1.6 mm, respectively). (**f**) Petri plate with callus clumps suitable for *Agrobacterium*-mediated transformation (bar = 16 mm). (**g**) Transient GFP expression in seedling-derived callus; 2 days after inoculation (bar = 1.7 mm). (**h**) Regeneration of GFP positive shoots on MSGR$^-$ medium (bar = 4 mm). (**i**) Rooting of plant in Phytatray™ with MSGR$^-$ medium (bar = 16 mm).

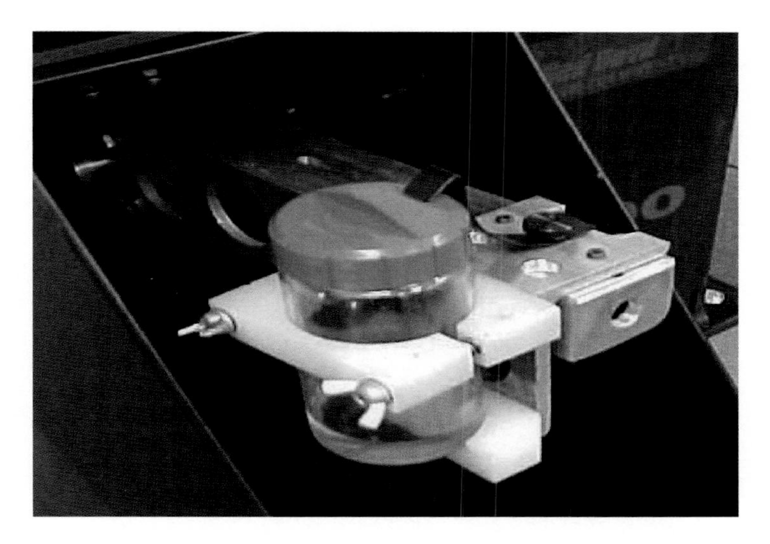

Chapter 5, Fig. 1. Closeup of modified Red Devil paint mixer with centrifuge bottle attachment.

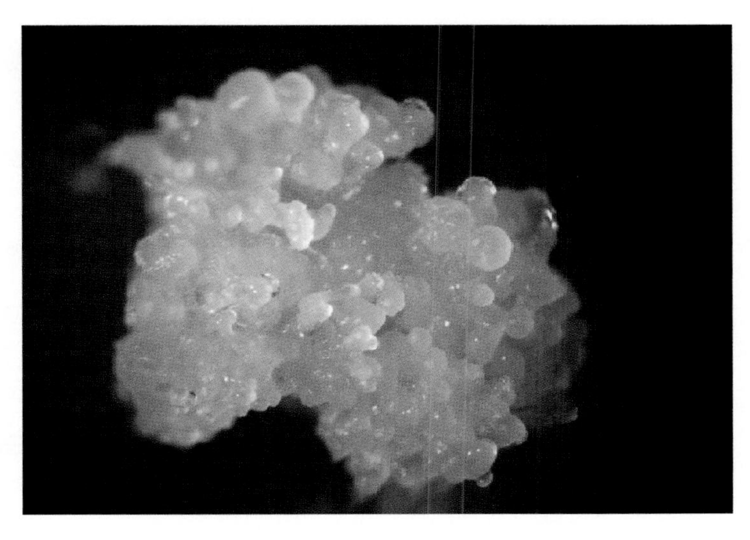

Chapter 5, Fig. 2. Friable embryogenic callus forming on the scutellar surface of immature embryos 2 weeks after culture on "15Ag10" medium.

Chapter 5, Fig. 3. Closeup of whiskers-treated tissue collected using a glass filter holder and filter flask connected to a house vacuum line.

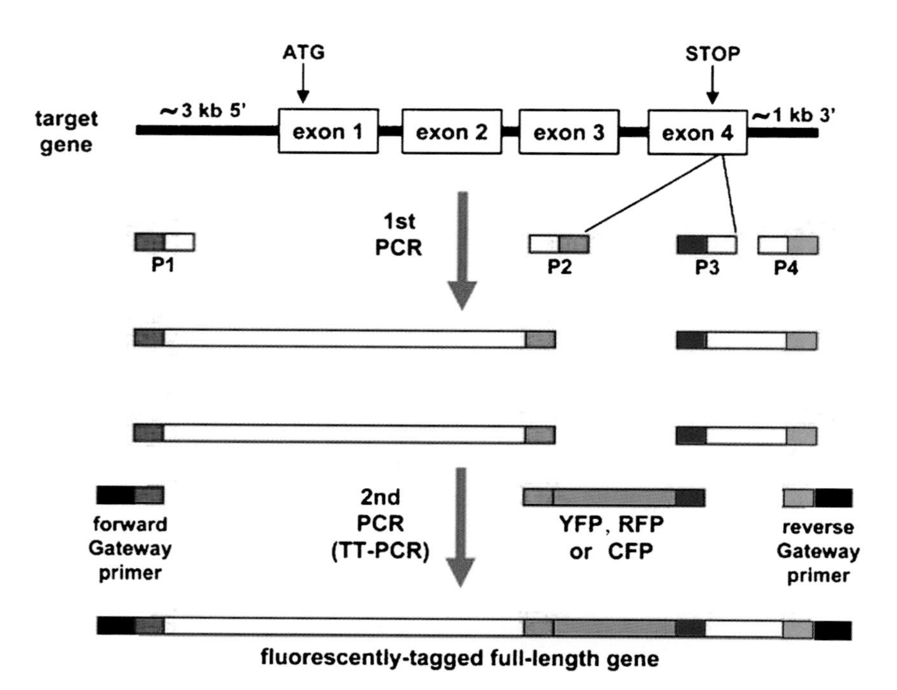

Chapter 6, Fig. 1. Schematic of the TTPCR method that is used to tag genes with fluorescent proteins for expression using their native regulatory sequences. Primers are designated P1-P4 and reverse/forward gateway, as described in the methods. Figure is adapted from *(9)*.

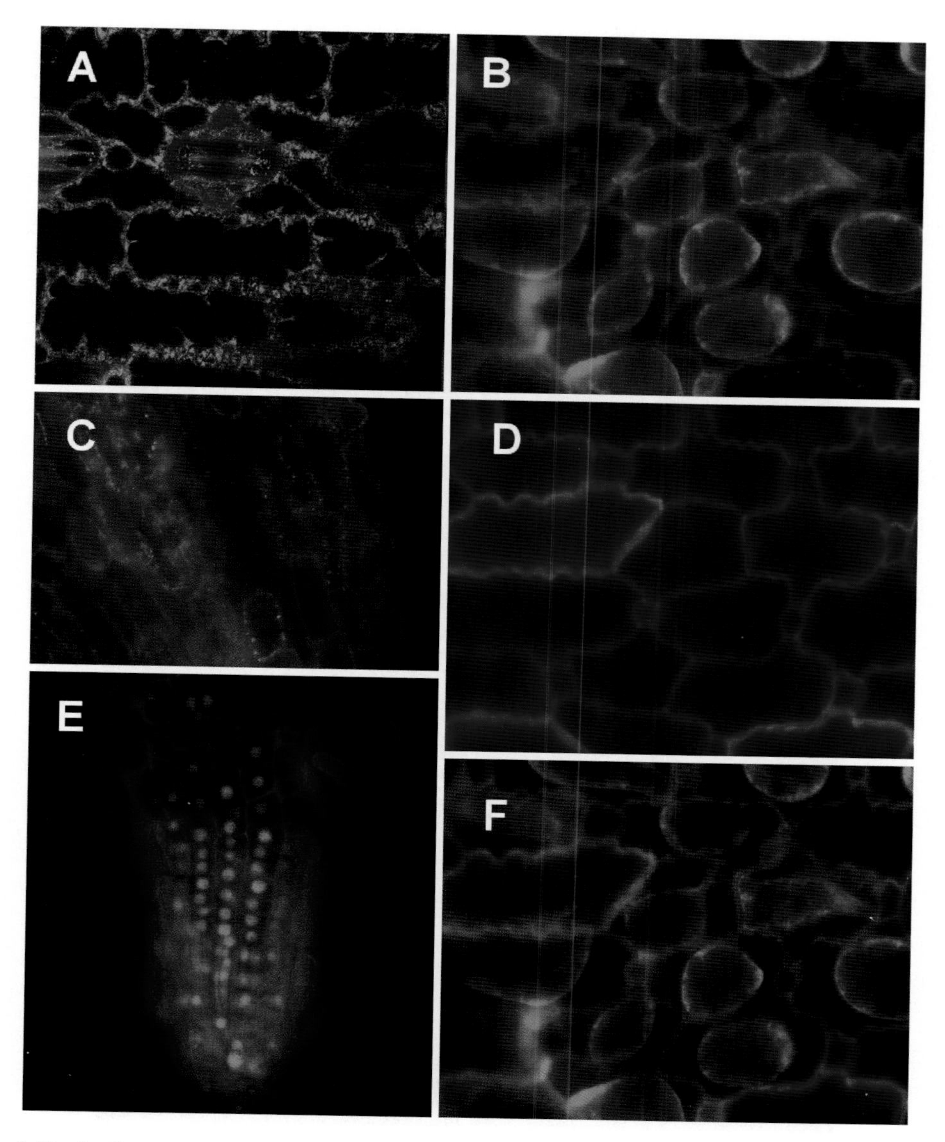

Chapter 6, Fig. 2. Expression of fluorescent protein fusions in maize using wide-field microscopy. (a) Localization of YFP-ExpansinA1 in cytoplasm and wall space of hand paradermal sections of maize leaves, counterstained with PI. (**b**, **d**, **f**) Merging of images identify wall and cytoplasmic localization in plasmolyzed cells of YFP-ExpansinA1 in juvenile leaf cells [(**b**) YFP channel, (**d**) PI channel, (**f**) merged YFP and PI channels, AS and AL, unpublished]. (**c**) Localization of YFP tagged HSP22 to mitochondria in expanding root cells, merged with UV light-induced cell wall autofluorescence. Note punctate green fluorescence typical of mitochondrial distribution (AS and AL, unpublished). (**e**) Localization of YFP fusion of a DNA repair protein to nuclei of root tip cells, highlighted by merging with wall autofluorescence (AS, AL, and Cliff Weil, unpublished).

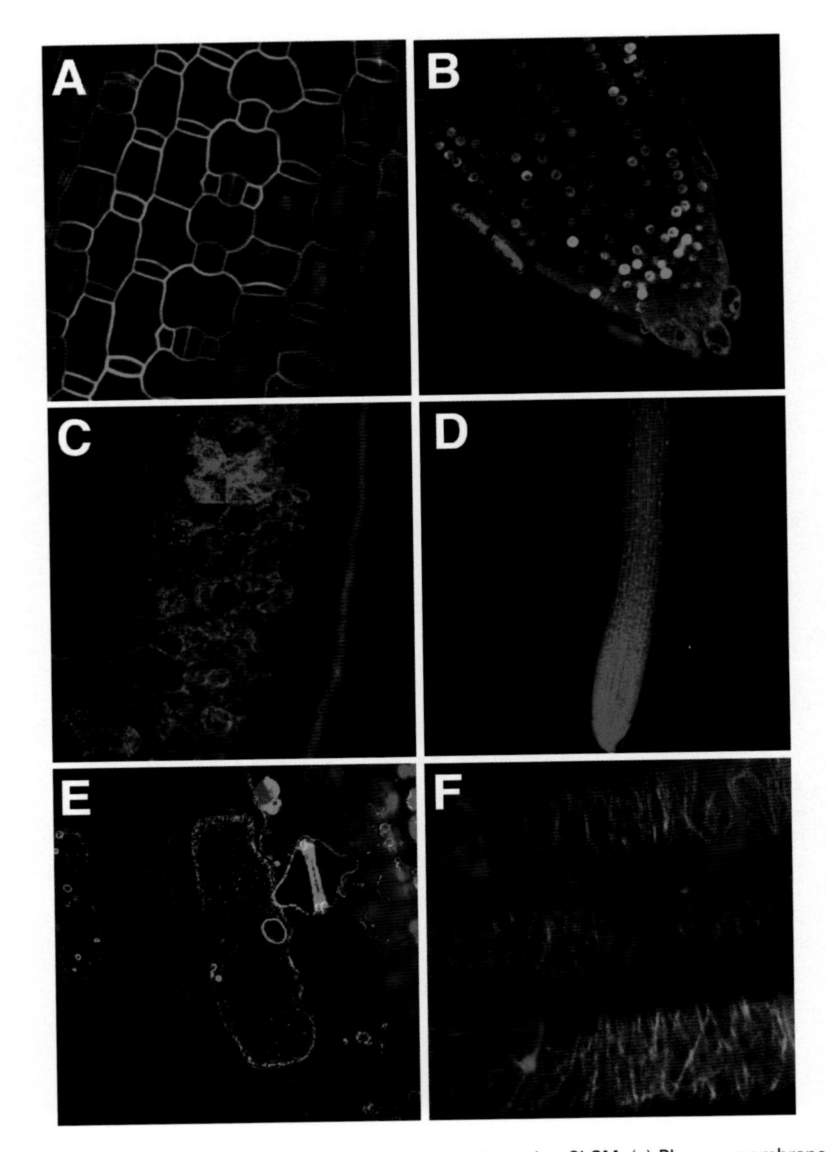

Chapter 6, Fig. 3. Visualization of fluorescent protein fusions in maize using CLSM. (**a**) Plasma-membrane localization of a PINFORMED1-YFP fusion in maize leaf (YY, Andrea Gallavotti and DJ, unpublished). (**b**) Nuclear localization of a Histone H1-YFP fusion (in *green*) in root tip, cells are counterstained using FM4-64 dye (molecular probes) (AM and DJ, unpublished). (**c**) Expression of a zein-RFP fusion in developing endosperm tissues, note the punctate localization corresponding to protein bodies. Autofluorescence of the aleurone layer is in *green* (YY and DJ, unpublished). (**d**) Expression of a maize histidine phosphotransfer protein-RFP fusion in the root (AM and DJ, unpublished). (**e**) Localization of a tonoplast intrinsic protein-YFP fusion in membranes, presumably vacuolar, of leaf cells (AM and DJ, unpublished). (**f**) Expression of a tubulin-YFP fusion labels microtubules in maize root cells. Plasma membranes are counterstained using FM4-64 dye (molecular probes) (AM, AS and DJ, unpublished).

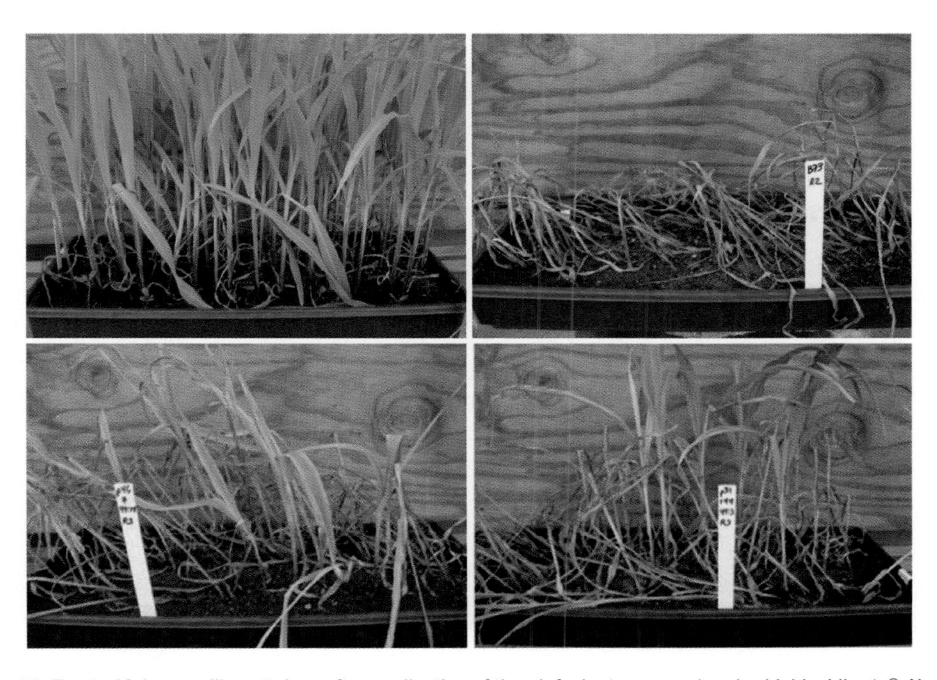

Chapter 12, Fig. 1. Maize seedlings 7 days after application of the glufosinate ammonium herbicide, Liberty®. Nontransgenic seedlings shown in *upper left* received no herbicide treatment, nontransgenic seedlings shown in *upper right* photo received the same treatment as seedlings of transgenic seedlots shown in *lower left* and *lower right*. Seedlings resistant to the herbicide shown for the transgenic seedlots were scored positive for the *Bar* gene. All conditions were as those outlined in **Subheading 3**.

Chapter 12, Fig. 2. Mature maize leaves from two different plants in the field treated with herbicide solution as described in **Note 2**. *Upper* leaf in *right hand* shows no symptoms of herbicide damage whereas *lower* leaf in *left hand* shows severe damage.

layer and all mesophyll cells, leaving the exposed lower epidermal layer. Careful rinsing releases loose cells, debris, and autofluorescing chloroplasts.

(d) Dissection: Most dividing and expanding cells are accessible only by removal of outer expanded portions of the plant. Developing ears (2–10 mm) can be dissected at ~6–8 weeks after planting. Dividing zones of leaf blades can be obtained by removing all outer leaves to access a zone about 1 cm above the ligule, when the sheath is ~1 cm long. Further dissection of the plant will allow isolation of the shoot apex for identifying meristem-specific localization.

(e) Counterstains and experimental manipulation are essential to help reveal protein location and to confirm normal function of the tagged protein. Counterstaining allows the visualization of cellular structures not labeled by the FP, for example, propidium iodide (PI) serves as an excellent *in vivo* stain to confirm cell viability as well as providing wall counterstaining (**Fig. 2**) and FM4-64 can be used to stain plasma membranes (**Fig. 3**). Experimental manipulation, such as inducing reversible plasmolysis, can distinguish cell wall from cortical cytoplasmic localization (**Fig. 2**). Selective experimentation and counterstaining will help resolve finer levels of detail, enhancing the information available for function of FP-tagged proteins.

4. Notes

1. Maize BAC clones are available from the Children's Hospital Oakland Research Institute, CHORI, http://bacpac.chori.org/, or the Arizona Genomics Institute, http://www.genome.arizona.edu/ or Clemson University Genomics Institute, http://www.genome.clemson.edu/.
2. Take care not to expose the fragments to short-wave UV light, which can cause damage. Either a long-wave UV light can be used, or the DNA fragments in the gel can be stained using methylene blue to avoid damage.
3. Because the *ccdB* gene present in pDONR 207 is toxic to most commonly used bacterial cells such as DH10B, DH5a, etc., the plasmid should be propagated in DB3.1 cells, which carry the *gyrA462* gene conferring resistance against *ccdB*. After the BP reaction, the *ccdB* gene is replaced by the TTPCR product; hence, the positive clones containing the TTPCR product can be maintained in standard *E. coli* strains such as DH10B, DH5α, etc.

4. BP and LR enzymes are stored in the –80°C freezer and thawed just before use. Mix gently by tapping the tube with your finger or mild vortexing.
5. In our hands, some of the unrecombined pDONR207 is able to grow in DH10B or DH5a cells. However, the number of colonies obtained from such unrecombined plasmid is approximately 10 times lower than that obtained in DB3.1 cells. Often the sizes of the colonies vary. Most probably, the smaller ones result from unrecombined vector. Choose medium to large colonies (large colonies may contain smaller, truncated inserts).
6. The yield and quality of plasmid DNA from *Agrobacterium* is usually low; this can be improved by incorporating a phenol-chloroform extraction in the mini prep method, or the plasmids can be transformed back into *E. coli*, then isolated, and used for restriction analysis.

Acknowledgments

We thank members of our maize GFP tagging project for useful discussions. We acknowledge the Iowa State University Plant Transformation Facility for providing excellent transformation services to the public sector. Our research is funded by National Science Foundation Grant DBI # 0501862 to Dave Jackson, Anne Sylvester, and Agnes Chan.

References

1. Chalfie, M., Tu, Y., Euskirchen, G., Ward William, W. and Prasher Douglas, C. (1994) Green fluorescent protein as a marker for gene expression. *Science* 263 802–805.
2. Shaner, N. C., Steinbach, P. A. and Tsien, R. Y. (2005) A guide to choosing fluorescent proteins. *Nat. Meth.* 2(12) 905–909.
3. Mathur, J., (2007) The illuminated plant cell. *Trends Plant Sci.* 12(11) 506–513.
4. Patterson, G. H. and Lippincott-Schwartz, J. (2002) A photoactivatable GFP for selective photolabeling of proteins and cells. *Science.* 297(5588) 1873–1877.
5. Griesbeck, O., Baird, G. S., Campbell, R. E., Zacharias, D. A. and Tsien, R. Y. (2001) Reducing the environmental sensitivity of yellow fluorescent protein. Mechanism and applications. *J. Biol. Chem.* 276 29188–29194.
6. Boute, N., Jockers, R. and Issad, T. (2002) The use of resonance energy transfer in high-throughput screening: Bret versus fret. *Trends Pharmacol. Sci.* 23 351–354.
7. Cutler, S. R., Ehrhardt, D. W., Griffitts, J. S. and Somerville, C. R. (2000) Random gfp::Cdna fusions enable visualization of subcellular structures in cells of arabidopsis at a high frequency. *Proc. Natl. Acad. Sci.USA.* 97 3718–3723.
8. Koroleva, O. A., Tomlinson, M. L., Leader, D., Shaw, P. and Doonan, J. H. (2005) High-throughput protein localization in arabidopsis using *Agrobacterium*-mediated transient expression of gfp-orf fusions. *Plant J.* 41 162–174.
9. Tian, G.-W., Mohanty, A., Chary, S. N., Li, S., Paap, B., Drakakaki, G., Kopec, C. D., Li, J.,Ehrhardt, D., Jackson, D., Rhee, S. Y., Raikhel, N. V. and Citovsky, V. (2004) High-throughput fluorescent tagging of full-length arabidopsis gene products in planta. *Plant Physiol.* 135 25–38.

10. Williams, P., Hardeman, K., Fowler, J. and Rivin, C. (2006) Divergence of duplicated genes in maize: Evolution of contrasting targeting information for enzymes in the porphyrin pathway. *Plant J.* 45 727–739.
11. Saleh, A., Lumbreras, V., Lopez, C., Kizis, E. D. P.-D. and Pages, M. (2006) Maize dbf1-interactor protein 1 containing an r3h domain is a potential regulator of dbf1 activity in stress responses. *Plant J.* 46 747–757.
12. Herrmann, M. M., Pinto, S., Kluth, J., Wienand, U. and Lorbiecke, R. (2006) The PTI1-like kinase ZmPti1a from maize (*Zea mays* l.) co-localizes with callose at the plasma membrane of pollen and facilitates a competitive advantage to the male gametophyte. *BMC Plant Biol.* 6 22.
13. Marton, M. L., Cordts, S., Broadhvest, J. and Dresselhaus, T. (2005) Micropylar pollen tube guidance by egg apparatus 1 of maize. *Science* 307 573–576.
14. Dresselhaus, T., Amien, S., Marton, M., Strecke, A., Brettschneider, R. and Cordts, S. (2005) Transparent leaf area1 encodes a secreted proteolipid required for anther maturation, morphogenesis, and differentiation during leaf development in maize. *Plant Cell.* 17 730–745.
15. Ma, Z. and Dooner, H. K. (2004) A mutation in the nuclear-encoded plastid ribosomal protein s9 leads to early embryo lethality in maize. *Plant J.* 37 92–103.
16. Ono, A., Kim, S.-H. and Walbot, V. (2002) Subcellular localization of mura and murb proteins encoded by the maize mudr transposon. *Plant Mol. Biol.* 50 599–611.
17. Beardslee, T. A., Roy-Chowdhury, S., Jaiswal, P., Buhot, L., Lerbs-Mache, S., Stern, D. B. and Allison, L. A. (2002) A nucleus-encoded maize protein with sigma factor activity accumulates in mitochondria and chloroplasts. *Plant J.* 31 199–209.
18. Taguchi-Shiobara, F., Yuan, Z., Hake, S. and Jackson, D. (2001) The fasciated ear2 gene encodes a leucine-rich repeat receptor-like protein that regulates shoot meristem proliferation in maize. *Genes Dev.* 15 2755–2766.
19. Rottgers, K., Krohn, N. M., Lichota, J., Stemmer, C., Merkle, T. and Grasser, K. D. (2000) DNA-interactions and nuclear localisation of the chromosomal hmg domain protein ssrp1 from maize. *Plant J.* 23 395–405.
20. Ivanchenko, M., Vejlupkova, Z., Quatrano, R. S. and Fowler, J. E. (2000) Maize rop7 gtpase contains a unique, caax box-independent plasma membrane targeting signal. *Plant J.* 24 79–90.
21. Landy, A. (1989) Dynamic, structural, and regulatory aspects of lambda site-specific recombination. *Annu. Rev. Biochem.* 58 913–941.
22. Chan, A. P., Pertea, G., Cheung, F., Lee, D., Zheng, L., Whitelaw, C., Pontaroli, A. C., SanMiguel, P., Yuan, Y., Bennetzen, J., Barbazuk, W. B., Quackenbush, J. and Rabinowicz, P. D. (2006) The tigr maize database. *Nucl. Acids Res.* 34 D771–776.
23. Fu, Y., Emrich, S. J., Guo, L., Wen, T. J., Ashlock, D. A., Aluru, S. and Schnable, P. S. (2005) Quality assessment of maize assembled genomic islands (magis) and large-scale experimental verification of predicted genes. *Proc. Natl. Acad. Sci. USA* 102 12282–12287.
24. Wach, A. (1996) PCR-synthesis of marker cassettes with long flanking homology regions for gene disruptions in. *S. cerevisiae. Yeast* 12 259–265.
25. Rozen, S. and H. Skaletsky (1999) Primer3 on the WWW for general users and for biologist programmers. *Meth. Mol. Biol.* 132 365–386.
26. Campbell, R. E., Tour, O., Palmer, A. E., Steinbach, P. A., Baird, G. S., Zacharias, D. A. and Tsien, R. Y. (2002) A monomeric red fluorescent protein. *Proc. Natl. Acad. Sci. USA.* 99 7877–7882.
27. Paz, M., Martinez, J., Kalvig, A., Fonger, T. and Wang, K. (2006) Improved cotyledonary node method using an alternative explant derived from mature seed for efficient *Agrobacterium*-mediated soybean transformation. *Plant Cell Rep.* 25 206–213.

Chapter 7

Use of Transgene-Induced RNAi to Regulate Endogenous Gene Expression

Karen McGinnis

Summary

RNAi can be an effective means to regulate endogenous gene expression in maize and as such represents an important reverse genetics tool. This approach involves designing a transgenic construct that creates a double stranded RNA (dsRNA) upon transcription, and introducing the transgene into maize plants. Transgenic lines bearing such a construct can be generated and characterized that are deficient in the gene of interest. Some variability has been observed in the efficiency of this technique, and there are several important aspects to consider. Herein, a basic protocol for using transgene-induced RNAi in maize is described, and some important considerations that can influence the success of this approach are discussed.

Keywords: RNAi, siRNA, Transgene, Gene silencing, Maize, dsRNA, Inverted repeat

1. Introduction

In plants and many other organisms, double stranded RNA molecules can induce transcriptional or posttranscriptional silencing of genes that share homology with the inducing RNA [reviewed in *(1)*]. This process is referred to as RNA-interference, or RNAi, and is dependent on the activity of several pathway specific proteins (**Fig. 1**). RNAi is an important regulator of endogenous gene expression and is critical to many aspects of plant development [reviewed in *(2)*]. RNAi-mediated silencing can also be utilized as a reverse-genetics tool, by creating a transgene that contains a portion of the endogenous gene in an inverted repeat

M. Paul Scott (ed.), *Methods in Molecular Biology: Transgenic Maize, vol. 526*

DOI: 10.1007/978-1-59745-494-0_7

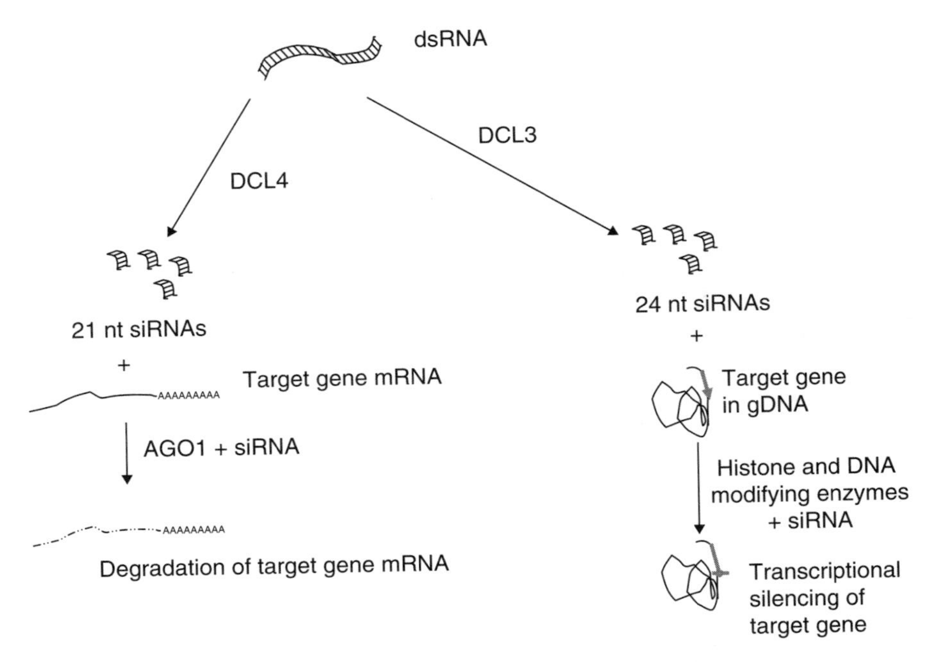

Fig. 1. Overview of RNAi pathways in plants. Where specific gene names are mentioned, *Arabidopsis* nomenclature was used, because gene function was genetically or biochemically determined in *Arabidopsis*.

orientation, such that transcription of the transgene results in a double stranded RNA [**Fig. 2**; *(3)*; reviewed in *(4)*].

An RNAi-mediated approach has been exploited to study gene function in both monocots and dicots *(5–10)*, with varying efficiencies. The basic principle behind this approach is to create a transgenic construct with a portion of the gene of interest cloned in an inverted repeat orientation around a spacer sequence such that transcription will result in two complementary sequences on the same molecule with an intervening spacer sequence to facilitate folding into a dsRNA (**Fig. 2**). To induce posttranscriptional silencing (mRNA degradation), the included sequence is typically homologous to a portion of the coding region of the gene to be silenced *(3, 9)*, while inclusion of a sequence homologous to the promoter region of the gene to be studied will induce transcriptional silencing (inhibition of transcription) of the targeted gene *(10)*.

In addition to the selection of the target gene sequence for inclusion in the inverted repeat construct, there are several other important factors to consider for optimal silencing of a given gene. These include construct design, target gene expression pattern, and measurement of target gene silencing efficiency. These features are important to ensure that the target gene is effectively silenced at the physiologically relevant stage(s) and tissue(s). A robust phenotype can provide an excellent way to identify

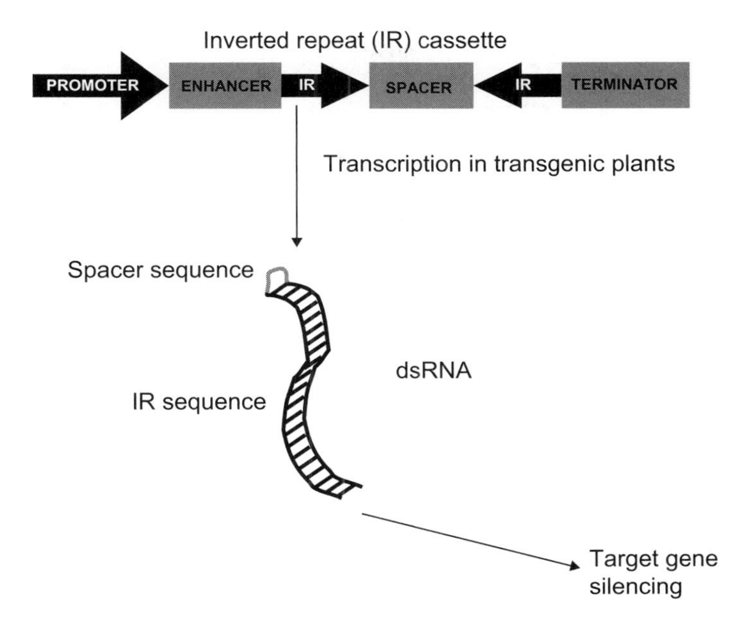

Fig. 2. Overview of transgene-induced RNAi mechanism in plants.

lines in which silencing has been the most effective, but such an indicator is not always available. In the absence of a predicted phenotype, target protein or mRNA levels can be measured to determine when silencing has occurred. One such method involves comparison of target gene mRNA levels to a uniformly expressed control gene *(7)*. Herein, these and other critical features are discussed in the context of a protocol for generating an inverted repeat-transgenic construct for inducing silencing of an endogenous gene in maize (summarized in **Fig. 3**).

2. Materials

2.1. Polymerase Chain Reaction of Inverted Repeat Sequence

1. Template DNA for amplification, which can be cDNA generated from total RNA extracted from a tissue or developmental stage where the gene is expressed, a cDNA clone of the gene of interest, or genomic DNA if promoter sequence is to be used.
2. Forward and reverse primers that will amplify the gene to be silenced and that include cloning adaptor sequences on the 5′ends so that the necessary restriction sites will be incorporated into the amplified product (*see* **Note 1**; **Fig. 3**). Stocks

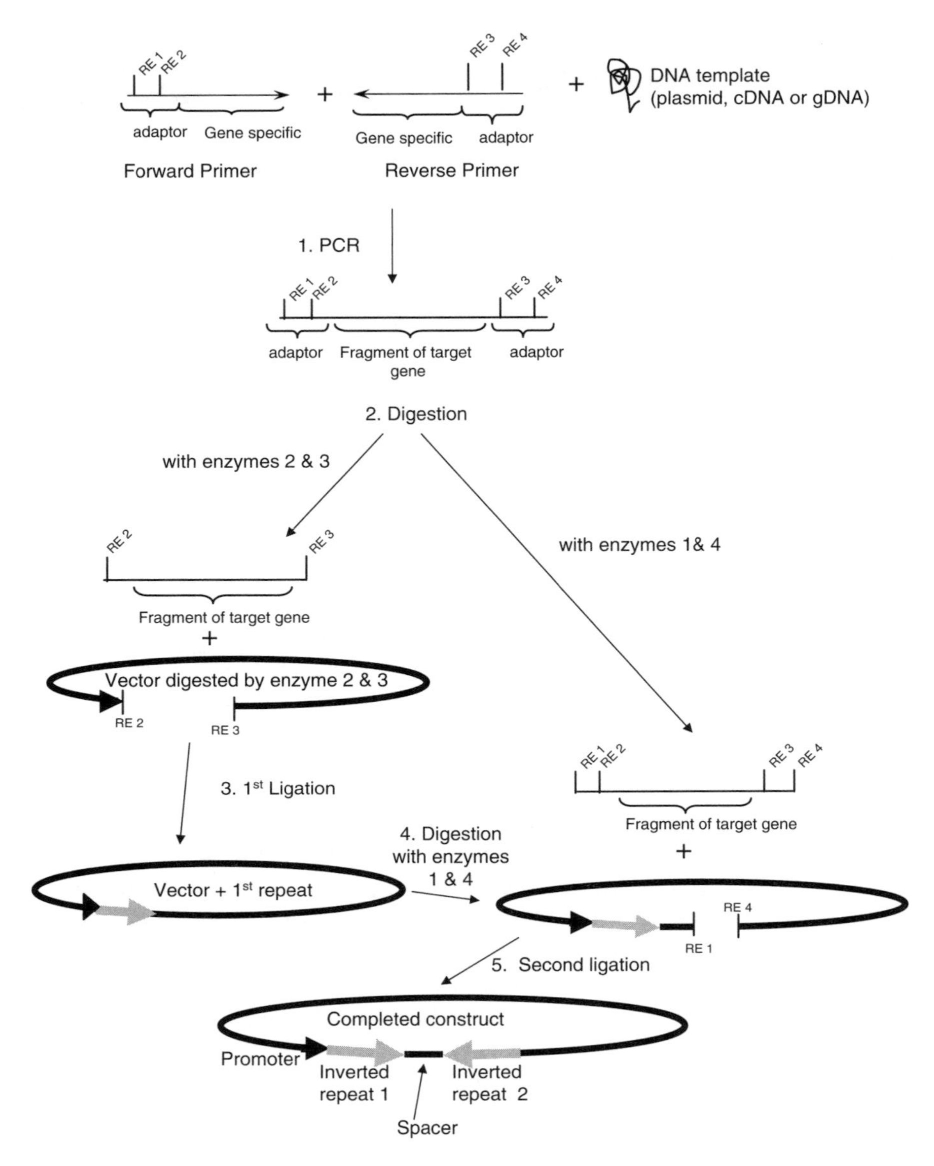

Fig. 3. Generalized two-step cloning strategy to generate inverted repeat transgenes (RE = restriction enzyme recognition site).

can be stored at 100 μM in water. For use, dilutions can be made in water to a concentration of 10 μM for each primer, and the dilutions combined to make a primer cocktail with a concentration of 5 μM per primer.

3. Polymerase chain reaction (PCR) reagents from manufacturer of choice: Taq polymerase, buffer (with magnesium chloride or add separately, according to manufacturer's instructions), dNTPs.

4. Purification kit for PCR product, such as QIAQuick PCR Purification Kit (Cat. # 2814, Qiagen, Valencia, CA).

2.2. Two-Step Cloning of Inverted Repeat Sequence

1. Restriction enzymes and recommended buffers (will need four restriction enzymes, identity of which depend upon adaptors in PCR primers).
2. Cloning vector: Can design custom vector to suit purposes (*see* **Note 2**), or use a publicly available vector, such as pMCG161 *(3)*, which can be ordered through ABRC (http://www.arabidopsis.org, stock number CD3–459).
3. DNA Ligase and appropriate buffer.
4. Millipore 0.02-μM pore size membranes (Millipore Corporation, Billicera, MA) for desalting ligations.
5. Alkaline phosphatase and buffer.
6. Electrocompetent cells for transforming ligated vectors.
7. SOB agar plates with appropriate antibiotic for selection of transformed colonies.

3. Methods

3.1. Polymerase Chain Reaction of Inverted Repeat Sequence

1. Amplify PCR product from EST plasmid or genomic DNA using gene specific primers with adaptors with standard molecular biological techniques. Perform two 100 mL reactions.
2. Analyze a 5-μL aliqout of each PCR product on an agarose gel to ensure that a single band of the predicted size is observed.
3. If PCR products appear to be the predicted size, follow manufacturer instructions to purify the PCR reactions with PCR clean up kit.

3.2. Two Step Cloning of Inverted Repeat Sequence

1. Digest the cloning vector (10 μL) and the PCR product (10 μL) with the innermost restriction enzymes (RE 2 and 3 in **Fig. 3**). Use the appropriate buffer and digestion conditions, as described by the restriction enzyme manufacturer.
2. Dephosphorylate the vector with alkaline phosphatase to enhance cloning efficiency.
3. Follow manufacturer instructions to deactivate the restriction enzymes and SAP by heating, freezing, or protein extraction.
4. Check digestion of vector and relative concentration of PCR product and vector by visualizing a 1-μL aliqout on an agarose gel.
5. Ligate vector with PCR product according to ligase manufacturer's instructions (*see* **Note 3**).

6. Desalt ligations using Millipore nylon disks. Cut disks into four parts. Label ligation number on disk (shiny side up.) Fill petri dish with nanopure water. Put disk on water surface. Apply ligation as a single drop on the middle of the disk. Cover the plate to avoid evaporation. Incubate for 1 h at room temperature. Collect desalted ligation in a new tube.
7. Transform 5 μl of ligation into electrocompetent *E. coli* cells by electroporation.
8. Incubate transformed cells in SOC for 1 h at 37°C
9. Plate all cells on SOB agar plates with selectable marker (chloramphenicol if using pMCG161 as a vector). Grow at 37°C overnight or until colonies are visible.
10. Streak out eight colonies on SOB agar plates with antibiotic for selectable marker.
11. Extract plasmid DNA from *E. coli* cells using standard protocols.
12. Digest plasmid DNA and visualize by agarose gel electrophoresis and staining to confirm predicted digestion pattern based on the sequence of ligation product. If clone looks correct, confirm with an additional restriction digest with different enzymes.
13. Digest the ligation product from **step 5** and the original PCR product with the outermost restriction enzymes (RE 1 and 4 in **Fig. 3**). Use the appropriate buffer and digestion conditions, as described by the manufacturer for the restriction enzymes to be used.
14. Dephosphorylate the vector with alkaline phosphatase to enhance cloning efficiency.
15. Follow manufacturer instructions to deactivate the restriction enzymes and SAP by heating, freezing, or protein extraction.
16. Check digestion of vector and relative concentration of PCR product and vector by visualizing a 1-μL aliqout on an agarose gel.
17. Ligate vector with PCR product according to ligase manufacturer's instructions.
18. Desalt ligations using Millipore nylon disks. Cut disks into four parts. Label ligation number on disk (shiny side up.) Fill petri dish with nanopure water. Put disk on water surface. Apply ligation as a single drop on the middle of the disk. Cover the plate to avoid evaporation. Incubate for 1 h at room temperature. Collect desalted ligation in a new tube.
19. Transform 5 μl of desalted ligation into electrocompetent *E.coli* cells by electroporation.

20. Extract plasmid DNA from *E. coli* cultures and confirm identity of clones by digestion and sequencing. Once the desired clone has been identified, the clone can be used to generate transgenic plants, according to the protocol described in a separate chapter of this volume or any other transformation service or protocol.

3.3. Identification of Lines with Robust Target Gene Silencing

1. Following identification of transgenic plants, outcross transgenic T_0 individuals with wild type, nontransgenic genotype of choice (*see* **Note 4**). Plant seed, allow to germinate, and collect tissue from T_1 individuals.
2. Plants from the T_1 generation can be analyzed by herbicide resistance or southern blot analysis to determine transgene presence, cosegregation, intactness, and insertion number. Target gene silencing can be assayed in this generation, or outcrossing can be repeated in this generation and analysis for silencing can be performed on later generations (*see* **Note 5**).
3. For analysis of target gene silencing, collect tissue from the plants to be assayed (*see* **Note 6**). Target gene silencing can be measured by RT-PCR analysis to determine mRNA abundance *(3, 7, 9)* or by analysis of protein abundance if a suitable antibody is available.

4. Notes

1. There are several important considerations for choosing the gene sequence to include in the inverted repeat. Because RNAi is a homology-dependent process, the sequence included in the inverted repeat will influence which, if any, related genes are silenced by the inverted repeat in addition to the primary target. This is of particular importance in maize, with its highly redundant genome. In some cases, it may be desirable to target both paralogs or multiple members of a gene family. Individualized targeting of related genes may be preferable in cases where silencing of more than one paralog or gene family member may be lethal, or when the purpose of the experiment is to determine the individual effect of a single gene. The included sequence will also influence the type of silencing that is induced, where cloning the promoter sequence in as the inverted repeat will induce transcriptional silencing *(10)*, and cloning a portion of the coding sequence (100–800 bp of sequence has been shown to be effective at inducing silencing of many genes) will induce posttranscriptional silencing *(9)*. It is crucial that the amplified sequence does not include any of the restriction sites that will be used for cloning in later steps, so the predicted PCR product should be carefully scanned for the presence of these sites.

2. Although there are vectors available for producing IR constructs, in some cases it may be optimal to design a custom vector that is suitable for silencing of specific genes. One important component of the vector that may warrant modifying an existing vector or building a custom vector is the promoter to drive expression of the inverted repeat. An adequate promoter is necessary to ensure production of sufficient double stranded RNA to induce efficient silencing of the endogenous target; as such a strong promoter is typically chosen for these experiments. For effective posttranscriptional silencing, the siRNAs must be present in the same tissue as the endogenous mRNA to effectively induce degradation of the mRNA. This requires that the promoter drives expression in overlapping tissues and developmental stages with the gene of interest. One strategy for achieving this is to use the defined promoter from the endogenous gene to be silenced to drive expression of the inverted repeat and therefore accomplish identical specificity; another strategy is to use a broadly expressed promoter whose expression pattern would include the relevant tissues and stages. For induction of transcriptional silencing, the promoter needs to result in siRNA production at a stage or tissue that completely overlaps with or precedes the expression of the target gene. If silencing is leaky, expression of the target gene may occur transiently and result in establishment of a physiological or a developmental condition that would not occur in complete absence of the target gene. Recent data suggests that an adequate terminator sequence is also crucial for minimizing posttranscriptional silencing of the transgenic construct *(11)*. In many cases, the plasmid backbone will influence the technique that can be used to introduce the transgenic construct into plants. As such, care should be taken to determine that a given backbone is *Agrobacterium*-compatible, if *Agrobacterium*-mediated transformation is to be used as a transformation method.
3. If possible, digest the ligation product with a restriction enzyme whose recognition sequence is found between the cloning sites in the uninserted vector. For example, in the example in **Fig. 3**, this would involve choosing an enzyme with a unique restriction site that lies anywhere between the two innermost (RE 2 and RE 3) of the restriction sites used for cloning the inverted repeats into. This will eliminate or minimize improperly digested or religated vector from the reaction. The cloned fragment from the target gene must be inspected to ensure that the restriction site is not also included in this sequence, however.
4. Existing data suggest that maintaining transgenic lines in a hemizygous condition increases the likelihood of transmitting

an active transgene faithfully into subsequent generations *(9)*. Thus, it is advisable to always increase or advance stocks by outcrossing with a nontransgenic genotype.

5. Silencing should be verified by RT-PCR or other method in any generation and tissue for which phenotypic analysis is relevant. Several generations of outcrossing will reduce the heterozygosity of genetic background, and may help reduce any epigenetic anomalies introduced during the transformation process, but the behavior of the transgene should be monitored throughout the generations. Assaying in early generations can guide the user in selecting a stable, silencing line, but continued observation is necessary in each new generation, because the transgene itself can become silenced or otherwise ineffective at inducing silencing of the target gene *(9)*. It cannot be assumed that an inverted repeat transgene is successfully inducing silencing at the T_4 generation, for example, because it did so in earlier generations of the same seed stock.
6. The optimal tissue for assaying silencing will be one in which the target gene is normally expressed, in addition to being relevant to the phenotype of interest, if possible.

References

1. Brodersen, P., and Voinnet, O. (2006) The diversity of RNA silencing pathways in plants. *Trends Genet* 22, 268–80.
2. Baurle, I., and Dean, C. (2006) The timing of developmental transitions in plants. *Cell* 125, 655–64.
3. McGinnis, K., Chandler, V., Cone, K., Kaeppler, H., Kaeppler, S., Kerschen, A., Pikaard, C., Richards, E., Sidorenko, L., Smith, T., Springer, N., and Wulan, T. (2005) Transgene-induced RNA interference as a tool for plant functional genomics. *Methods Enzymol* 392, 1–24.
4. Waterhouse, P. M., and Helliwell, C. A. (2003) Exploring plant genomes by RNA-induced gene silencing. *Nat Rev Genet* 4, 29–38.
5. Chen, J., Tang, W. H., Hong, M. M., and Wang, Z. Y. (2003) OsBP-73, a rice gene, encodes a novel DNA-binding protein with a SAP-like domain and its genetic interference by double-stranded RNA inhibits rice growth. *Plant Mol Biol* 52, 579–90.
6. Segal, G., Song, R., and Messing, J. (2003) A new opaque variant of maize by a single dominant RNA-interference-inducing transgene. *Genetics* 165, 387–97.
7. Kerschen, A., Napoli, C. A., Jorgensen, R. A., and Muller, A. E. (2004) Effectiveness of RNA interference in transgenic plants. *FEBS Lett* 566, 223–8.
8. Travella, S., Klimm, T. E., and Keller, B. (2006) RNA interference-based gene silencing as an efficient tool for functional genomics in hexaploid bread wheat. *Plant Physiol* 142, 6–20.
9. McGinnis, K., Murphy, N., Carlson, A. R., Akula, A., Akula, C., Basinger, H., Carlson, M., Hermanson, P., Kovacevic, N., McGill, M. A., Seshadri, V., Yoyokie, J., Cone, K., Kaeppler, H. F., Kaeppler, S. M., and Springer, N. M. (2007) Assessing the efficiency of RNA interference for maize functional genomics. *Plant Physiol* 143, 1441–51.
10. Cigan, A. M., Unger-Wallace, E., and Haug-Collet, K. (2005) Transcriptional gene silencing as a tool for uncovering gene function in maize. *Plant J* 43, 929–40.
11. Luo, Z., and Chen, Z. (2007) Improperly terminated, unpolyadenylated mRNA of sense transgenes is targeted by RDR6-mediated RNA silencing in *Arabidopsis*. *Plant Cell* 19, 943–58.

Chapter 8

Plasmid Rescue: Recovery of Flanking Genomic Sequences from Transgenic Transposon Insertion Sites

Guo-Ling Nan and Virginia Walbot

Summary

The transgenic *RescueMu* lines were designed for and successfully used in our maize gene discovery project. The pBluescript-containing *RescueMu* transposon can be readily recovered by a procedure called plasmid rescue. Plasmid rescue is a technique for recovering bacterial plasmids from transgenic eukaryotic genomic DNA. Total maize DNA was first digested with restriction enzyme(s), ligated, and then transformed into *E. coli* cells. Colonies were recovered under selection against the antibiotic marker(s) in the transgene vector. Ampicillin or carbenicillin was used for *RescueMu* transgene recovery. The flanking genomic sequences at *RescueMu* insertion sites were simultaneously captured and then sequenced using *RescueMu*-readout primers. Genomic DNA from an individual plant or from pooled samples of up to ~50 plants could be used in a single rescue. Because the majority of transgenic constructs currently used in flowering plants were made in the form of plasmids, this protocol could therefore be adapted by and useful to researchers involved in other transgenic work and be versatile for characterizing transgene loci.

Keywords: *RescueMu*, Transgene array, Electroporation, Colony lift

1. Introduction

Our gene tagging strategy, *RescueMu*, enables us to recover the genomic sequences flanking *RescueMu* insertion sites in plasmids that can be maintained in bacteria. Approximately 200,000 *RescueMu*-tagged maize genomic sequences, representing ~30,000 insertions or ~10 Mb predominantly in genic regions have been deposited in Genbank *(1)* and unpublished data). With the help of plasmid rescue (**Fig. 1**), we confirmed that *(1)* somatic as well as germinal insertions occurred; *(2)* germinal insertions were mostly late occurring; and *(3)* upon insertion, *RescueMu* generated the characteristic 9-bp insertion

M. Paul Scott (ed.), *Methods in Molecular Biology: Transgenic Maize, vol. 526*

DOI: 10.1007/978-1-59745-494-0_8

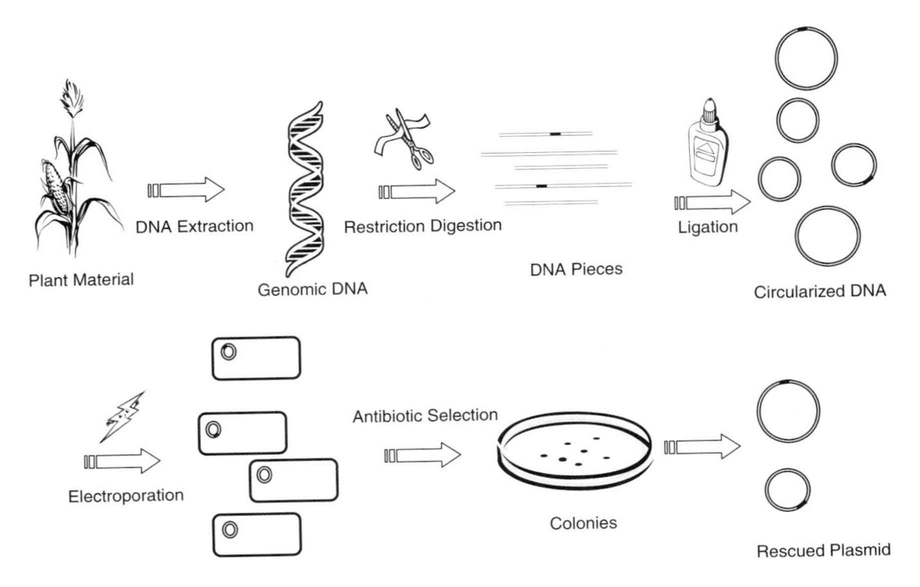

Fig. 1. Plasmid rescue procedure. *Double lines* are maize genomic DNA and *black solid lines* are transgene segments.

site duplication *(2)*. Although some insertion sites are easier to rescue than others for unidentified reasons, several experimental factors should be considered in optimizing the design for a high-quality rescue. Preferable restriction enzymes are those that will not cut inside the transgene or those that at least leave the plasmid portion functional for one-sided rescues. If two or more restriction enzymes are used, it is recommended that they have compatible ends. Ligation conditions should be set to favor self-circularization, to ensure that the plasmids are smaller in size without extra pieces. The most crucial measure in a successful rescue is eliminating/minimizing contamination, which could result from any commonly used plasmids with the same selection marker. Such contamination is a problem because of the low transformation efficiency of rescued plasmid compared to regular plasmids. The transformation efficiency of the rescued plasmids is much lower (100× less) than that of common plasmids. First, the relative plasmid concentration is low in plasmid rescue. There is only ~5 pg of single-copy transgene per μg of total genomic maize DNA in a hemizygous background. Second, the large (average over 10 kb) rescued plasmids contain foreign, eukaryotic sequence. Therefore, traces of contaminating common plasmids could easily give rise to an overwhelming number of unwanted colonies. To eliminate contamination we recommend working in a lab area dedicated to plasmid rescue and using dilute bleach or other rigorous clean-up procedures (UV-C lamps) before and after using the lab bench and equipment. Once a true rescue is obtained, one can design multiple sequencing primers going in the 5′ and/or 3′ directions to analyze the transgene locus. This robust plasmid rescue protocol allows us to recover not only the single-copy germinal insertion events but also the rare somatic insertion events that are found only in one or a sector of cells.

2. Materials

2.1. Plasmid-Free DNA Extraction

1. Urea extraction buffer (1 L): 7 M urea, 350 mM NaCl, 1% (w/v) *N*-lauroyl sarcosine, 50 mM Tris–HCl (pH 8.0), 20 mM EDTA (pH 8.0).
2. Charcoal, activated (Sigma, St. Louis, MO).
3. DNAZap™ (Ambion/Applied Biosystems, Austin, TX).
4. Buffer-saturated phenol:chloroform:isoamyl alcohol mix (25:24:1, v/v/v) (Invitrogen, Carlsbad, CA).
5. Chloroform:isoamyl alcohol mix (24:1, v/v).
6. DNase-, RNase-free distilled water, Molecular Biology Grade (Invitrogen, Carlsbad, CA).

2.2. Restriction Digestion and Ligation

1. Restriction enzymes, RNase T1 (Epicentre, Madison, WI) or RNase A (New England Biolabs, Beverly, MA).
2. T4 DNA Ligase (Invitrogen, Carlsbad, CA).
3. Molecular Biology Grade Low-EEO agarose (Fisher, Pittsburgh, PA).

2.3. Escherichia Coli Transformation

1. ElectroMax™ DH10B™*E. coli* competent cells (Invitrogen, Carlsbad, CA).
2. Electroporation cuvettes, 0.1 cm (Invitrogen, Carlsbad, CA).
3. S.O.C. medium (Invitrogen, Carlsbad, CA).
4. LB medium (1 L): 10 g bacto-tryptone, 5 g bacto-yeast extract, 10 g NaCl, adjusted to pH 7.2 and autoclaved (Benton and Dickinson, Sparks, MD).
5. LB Plates: 1 L LB medium + 12 g bacto-agar, autoclaved, and add 100 mg/L carbenicillin or ampicillin (1 mL of 100 mg/mL filter-sterilized stock).

2.4. Colony Lift Hybridization and Plasmid Extraction

1. Hybond™-N$^+$ positively charged nylon disc, 82 mm (GE Healthcare, Piscataway, NJ).
2. 10% (w/v) SDS solution.
3. 3MM filter paper (Whatman, Maidstone, UK).
4. Denaturation solution: 1.5 M NaCl, 0.5 M NaOH.
5. Neutralization buffer: 1.5 M NaCl, 0.5 M Tris–HCl (pH 7.5).

6. 20× SSC: 3 M NaCl, 0.3 M Na-citrate, pH 7.0.
7. AlkPhos Direct Labelling and Detection System (GE Healthcare, Piscataway, NJ).
8. CDP-Star™ detection reagent (GE Healthcare, Piscataway, NJ or New England Biolabs, Beverly, MA).
9. BioMax Light X-ray film (Eastman Kodak, Rochester, NY) or equivalent.
10. Qiaprep spin miniprep kit (Qiagen, Valencia, CA) or other plasmid purification cartridges.

3. Methods

3.1. Plasmid-Free DNA Extraction

3.1.1. Preparation of Plasmid-Free Reagents and Precaution

1. When possible, order ready-made buffers, solutions, solvents, and water from chemical suppliers.
2. Self-prepared solutions, e.g., urea buffer, should be treated with activated charcoal by adding a small amount of powder to the solution and mixing on a stirring plate for 5–10 min. The treated solution is then filtered through a 0.2-μm membrane.
3. Preclean work bench and cover it with fresh bench liners.
4. Preclean glass rods with DNAZap™ as directed by the manufacturer.
5. Use filter tips through steps until completion of electroporation (**Subheading 3.3.2**).
6. Follow the extraction procedure described in **Subheading 3.1.1** in Chapter 9.

3.1.2. Fluorescent DNA Quantification in 96-well Plates

1. Dilute Picogreen® dye (Invitrogen, Carlsbad, CA) 1:200 (v/v) with 1× TE buffer, making enough to assay all the samples and construct a standard curve (*see* **Note 1**).
2. Dispense 100, 95, 75, and 50 μL 1× TE into wells corresponding to 0, 10, 50, and 100 ng for the standard curve, respectively.
3. Dilute the DNA standard to 2 ng/μL and add 5 μL, 25 μL, and 50 μL of the diluted DNA standard into wells corresponding to 10 ng, 50 ng, and 100 ng, respectively.
4. Dispense 95 μL 1× TE into each sample well.
5. Dilute 5 μL of the genomic DNA sample in 20 μL 1× TE. Mix well and pipet 5 μL into the sample well.
6. Dispense 100 μL of the diluted Picogreen® from **step 1** into each well and mix by pipetting up and down several times (*see* **Note 2**).

7. Incubate for 3 min then quantify on a fluorescence microplate reader (excitation at 480 nm, emission at 520 nm). The original genomic DNA concentration generally falls between 100 and 300 ng/μL.

3.2. Restriction Digestion and Ligation

3.2.1. Restriction Digestion of Genomic DNA

1. Digest 10 μg genomic DNA in 400 μL 1× restriction buffer containing 50 units of each restriction enzyme plus either 500 units RNase T1 or 200 μg RNase A. Incubate for 1.5–3.0 h at the optimal digestion temperature.
2. In a fume hood, add 400 μL phenol:chloroform:isoamyl alcohol mix to each digested sample, close the cap and invert the tube vigorously for 1–2 min.
3. Centrifuge 5 min at 10,000 × *g* in a microcentrifuge.
4. In a fume hood with a P-1,000 pipet carefully transfer the upper aqueous phase to a fresh 2-mL tube.
5. Repeat **steps 2–4** with 400 μL chloroform:isoamyl alcohol mix.
6. Add 40 μL 3 M Na-acetate, mix well, add 1 mL 100% ethanol and mix again.
7. Incubate at 4°C or below for at least 10 min.
8. Centrifuge in a microcentrifuge at 10,000 × *g* for 20 min at 4°C.
9. Decant the ethanol supernatant.
10. Add 500 μL 70% ethanol at room temperature and mix by tapping the tube.
11. Centrifuge in a microcentrifuge at 10,000 × *g* for 1 min at room temperature.
12. Carefully decant the ethanol supernatant.
13. Repeat **steps 10–12** one more time.
14. Centrifuge again briefly to bring down any ethanol residue to the bottom of the tube.
15. Remove remaining ethanol carefully with a P-200 pipet.
16. Leave the cap open and position the opening of the tube facing the air flow in a laminar flow hood for 30 min or until the pellet is dried.

3.2.2. Ligation

1. Dissolve the digested DNA in a final volume of 400 μL containing: 1× ligation buffer (Invitrogen, Carlsbad, CA) and 10 units T4 DNA ligase (*see* **Note 3**).
2. Ligate for 1 h at room temperature or overnight at 4°C.
3. Repeat **steps 2–16** in **Subheading 3.2.1**.
4. Resuspend the ligated DNA in 10 μL Molecular Biology Grade distilled water.

5. Store on ice if used immediately. Store at –20°C or below if not used immediately.

3.3. E. Coli Transformation

3.3.1. Preparation for Electroporation

1. Thaw competent cells on wet ice (*see* **Note 4**).
2. Prechill electroporation cuvettes and 1.5-mL microcentrifuge tubes on wet ice.
3. Once the competent cells are thawed, carefully resuspend the cells and dispense 30 μL into each chilled microcentrifuge tube (*see* **Note 5**).

3.3.2. Electroporation

1. Add 1.5–2.0 μL of ligated DNA into the cells and gently mix with a pipet.
2. Transfer the DNA per cells mix into the 0.1-cm slot of a chilled cuvette (*see* **Note 6**).
3. Cover the cuvette and tap the cuvette firmly on a solid surface to bring the content to the bottom of the slot.
4. Electroporate at 100 Ω, 2.3–2.5 kV, and 25 μF (*see* **Note** 7).
5. Immediately add 1 mL S.O.C. medium to the cells, mix gently by pipetting and transfer to a culture tube.
6. Incubate at 37°C for 45–60 min on a rotary shaker at 200–250 rpm.
7. Plate aliquots of 200–350 μL cells directly on LB+ antibiotic plates; for pBluescript we used 100 mg/L carbenicillin.
8. Incubate at 37°C for 16–20 h.

3.4. Colony Lift Hybridization (Optional, see Note 8)

3.4.1. Colony Lift

1. Prechill plates with colonies at 4°C for at least 30 min.
2. Label a Hybond™-N+ nylon disc with a pencil, overlay on top of the agar and mark the disc position on the bottom of the plate.
3. After 30–60 s remove the disc from the plate using blunt-end forceps and keep the disc with the colony side up from now on through **step 10**.
4. Overlay the disc on 1–2 layers of 3MM paper soaked with 10% SDS for 2 min (lysis).
5. Blot dry on a sheet of 3MM paper.
6. Transfer onto 1–2 layers of 3MM paper soaked with denaturation solution for 3 min.
7. Blot dry on a sheet of 3MM paper.
8. Transfer onto 1–2 layers of 3MM paper soaked with neutralization buffer for 3 min.
9. Blot dry on a sheet of 3MM paper.
10. Repeat **steps 8** and **9** with a second batch of neutralization buffer and 3MM paper.
11. Rinse in excess amount of 2× SSC on a shaker with 2–3 changes of 2× SSC.

12. Dry the disc with the DNA side up on a sheet of 3MM paper at 65°C for 30 min.
13. Bake the disc at 80°C for 2 h (optional).

3.4.2. Membrane Hybridization and Signal Detection

1. Three to four discs can be simultaneously hybridized in a large hybridization tube containing 50 mL hybridization buffer.
2. Prehybridize at 65°C for at least 30 min on a rotary device in a hybridization oven.
3. Add 60–100 ng of labeled transgene-specific probe (preferably a segment outside the selection marker (*see* **Note 9**) – in the case of *RescueMu*, we use the 0.4-kb *Sinorhizobium* tag) and hybridize overnight at 65°C.
4. Washes and signal detection steps are as described in **Subheading 3.3.3** in Chapter 9.
5. X-ray film exposure time is usually less than 1 min.
6. Match and mark positive colonies on the LB plate.

3.5. Plasmid Extraction and Storage of Rescued Clones

1. Pick and grow individual colonies overnight in 3-mL fresh LB medium with 100 mg/L carbenicillin.
2. Extract plasmid DNA from 2 mL of overnight culture. Add 100% glycerol to the remaining culture to make 30% glycerol stock (for long-term storage in –80°C freezer).
3. Plasmid can be checked by restriction profiling (**Fig. 2**) prior to submission for sequencing.

4. Notes

1. Picogreen® is light sensitive and therefore, prolonged exposure to light should be avoided. It should be stored at –20°C in small aliquots to avoid repeated thawing.
2. Use a multichannel pipet or repeat pipet to minimize time lapse between samples for best results.
3. Use fresh ligation buffer for best results. If the mixture has been previously thawed, add 1 μL 10 mM ATP to the ligation mix.
4. The term "wet ice" is basically an ice bath filled with water for better immersion.
5. Avoid creating bubbles when pipetting or mixing the competent cells as bubbles tend to cause sparks during electroporation.
6. Tilt the cuvette and carefully dispense the DNA/cell mix along one side of the slot.

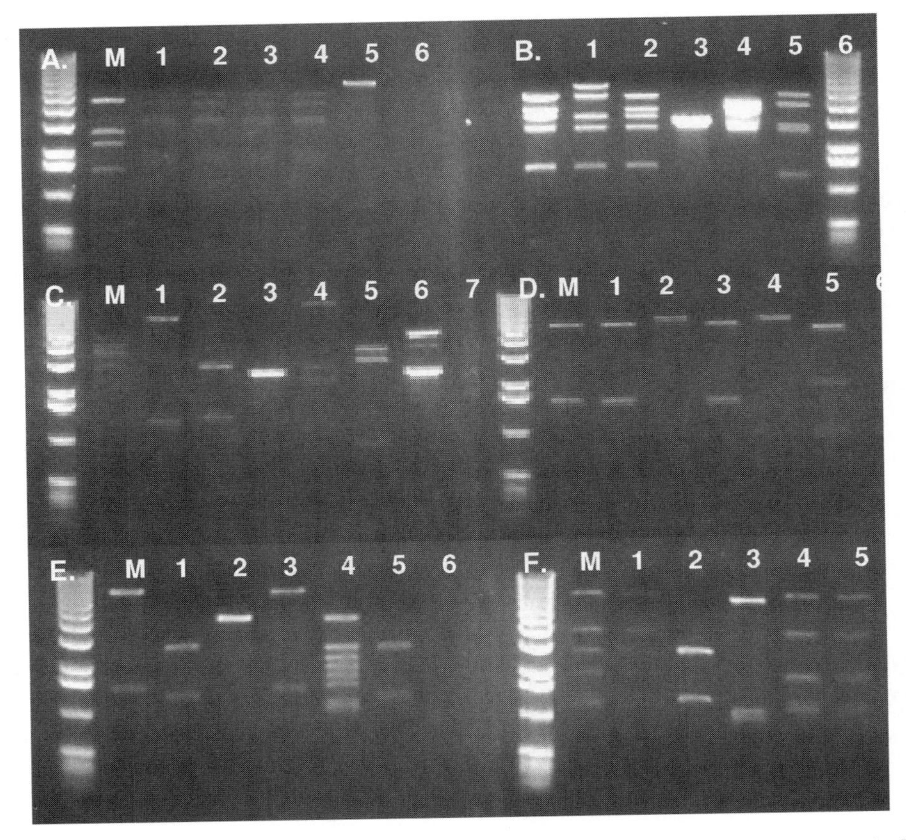

Fig. 2. Ethidium bromide-stained agarose gels showing restriction patterns of plasmids rescued from six *RescueMu* individuals (**a–f**) containing putative new insertion events in the original *RescueMu* transgene background. Genomic DNA was digested with *KpnI*, ligated, and digested with *BglII* to exclude the original arrays (*see* Fig. 1a in Chapter 9). After transformed into *E. coli*, positive clones were picked after colony hybridization with the *Sinorhizobium* probe. Purified plasmids were double-digested with *Eco*RI and *Hin*dIII and separated on 1% agarose gels. Clones (A2/A3/A4/A5,B1/B3, DA/D2/D4, E1/E4, F1/F2, and F5/F6) showing similar restriction profiles are likely germinal events and clones showing a distinct restriction profile could be either germinal or somatic events. M: 1 kb molecular weight marker (Invitrogen, Carlsbad, CA).

7. Excess salts or bad batches of cells also cause sparks. Reducing the amount of input DNA and/or the voltage usually reduces sparking, but the transformation efficiency is also compromised.
8. This step is only necessary when contamination is suspected. A good rescue, resulting from plasmids recovered from approximately 0.6 μg of starting genomic DNA, generally gives rise to a few hundred colonies of about 1 mm in diameter. When the number of colonies is too high and satellite colonies (tiny, secondary colonies surrounding a bigger one) are present, contamination is usually suspected and colony lift hybridization can be performed. The true rescues are sometimes distinguishable from the fake ones simply by the lack of satellite colonies around them.

9. Because the colonies have survived the selection antibiotic on the plate, it will not be adequate to use a probe from the selection marker gene for this step.

Acknowledgments

This project was supported by National Science Foundation Grant 98-72657 to V.W.

References

1. Fernandes, J., Dong, Q., Schneider, B., Morrow, D.J., Nan, G.-L., Brendel, V., and Walbot, V. (2004). Genome-wide mutagenesis of *Zea mays* L. using *RescueMu* transposons. *Genome Biol.* 5, R82.

2. Raizada, M.N., Nan, G.-L., and Walbot, V. (2001b). Somatic and germinal mobility of the *RescueMu* transposon in transgenic maize. *Plant Cell* 13, 1587–1608.

Part IV

Analysis of Transgenic Plants

Chapter 9

Nonradioactive Genomic DNA Blots for Detection of Low Abundant Sequences in Transgenic Maize

Guo-Ling Nan and Virginia Walbot

Summary

Sensitive and reproducible genotyping tools are fundamental in interpreting and substantiating genetic data. In cases where alternative assays like PCR are not applicable, a sensitive genomic Southern protocol is needed. Our maize gene discovery work using the *RescueMu* transgenic lines was such a task. The direct proof of each new germinal insertion event can be assessed only on a genomic DNA hybridization analysis, and therefore we developed the following protocol to screen efficiently through hundreds up to thousands of samples in a relatively short time. The DNA extraction protocol was scaled to accommodate samples processed in a microcentrifuge with consistent yield of ~50 μg of high molecular weight DNA. A trained person can easily process several hundred samples in a few days. Once the DNA is extracted, final results can be obtained routinely within a week on ~100 or more samples, depending on the capacity of the electrophoresis and hybridization apparatus available. Under our optimized conditions, the method described below generates blots with high sensitivity and low background even after repeated stripping and reprobing. Single to low-copy transgenes as well as maize genomic sequences can be detected consistently. The nonradioactive DNA probes employed are not only safer, compared to the conventional radioactive probes, but also greatly shorten the exposure time. Confident estimation of copy number – as good as quantitative PCR – and visualization of transgene complexity are just a few more advantages of this protocol.

Keywords: Transposon tagging, *RescueMu*, DNA blot hybridization, Non-radioactive probe

1. Introduction

MuDR/Mu transposons are characterized by a high forward mutation frequency (10^{-5} to 10^{-3} per locus) *(1)* and have been used in mutant screening through directed and random gene tagging. The *RescueMu* transposon was constructed to expedite the exploration of maize genomic sequences. A bacterial plasmid,

M. Paul Scott (ed.), *Methods in Molecular Biology: Transgenic Maize, vol. 526*

DOI: 10.1007/978-1-59745-494-0_9

pBluescript, was inserted in the center of a nonautonomous *Mu1* element to generate *RescueMu* (**Fig. 1a**). Insertion sites along with their flanking genomic sequences can be recovered in plasmid form and sequenced (detailed procedures for plasmid rescue are described in Chapter 8). To monitor the somatic activity of *RescueMu*, an excision reporter using the *Lc* gene (*R* gene family) required for anthocyanin pigment production was disrupted by *RescueMu*; when the transposon excises, the anthocyanin regulatory protein can be expressed (**Fig. 1a**). Biolistically transformed primary transgenic *RescueMu* lines (**Fig. 2**, T_0) contained complicated transgene arrays *(3)* in a non-Mutator HiII (A188 × B73) background. Active, autonomous *MuDR* elements were introduced through genetic crosses with various Mutator lines to mobilize *RescueMu*. Somatic excision sectors of one or a few cells were observed in an *r–r* or *r–g* background (**Fig. 1b**). The late timing was similar to other *Mu* excision events. While somatic transposition events can be visualized by inspecting kernels (**Fig. 2**, F_1), germinal insertion events, which are also developmentally late-occurring in the cells before meiosis, during meiosis, and in the haploid gametophytes, are scored by DNA blot hybridization in the next generation. Typically DNA was extracted from F_2 leaves (**Fig. 2**). Germinal transposition frequencies remained low (0–24% per plant) in families with the original, complex arrays *(2, 3)*. After several seasons of searching, a few plants containing

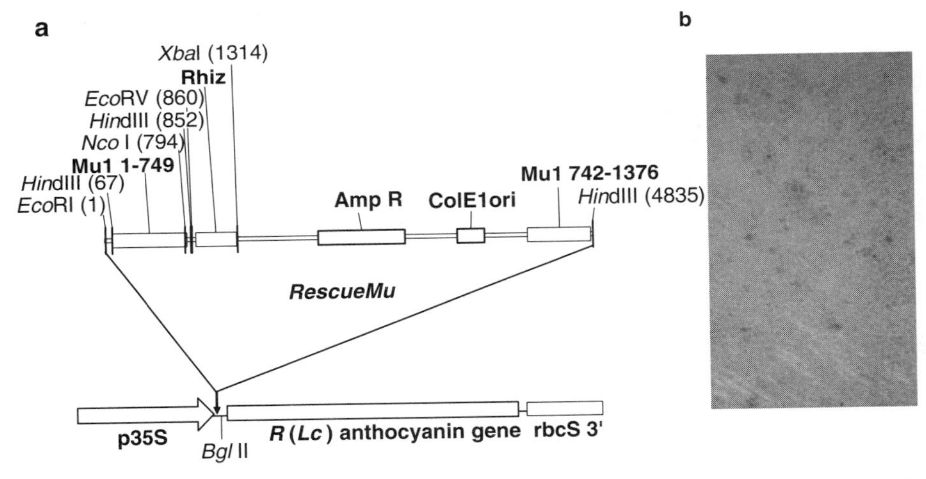

Fig. 1. (**a**) Diagram of the *RescueMu* vector including its somatic excision reporter *Lc* gene. The transposable element spans from the left TIR (terminal inverted repeat) to the right TIR. The complete sequence of the 4,739-bp *RescueMu* element was deposited in Genbank, accession AY301066. *Mu1* bases 1–749: 5′ half of the *Mu1* element, including the left TIR; *Mu1* 742–1376: 3′ half of the *Mu1* element, including the right TIR; Rhiz: a ~400-bp fragment from ***Sinorhizobium meliloti;*** Amp R: ampicillin resistance gene; ColE1: replication origin, p35S: the 35S promoter from Cauliflower Mosaic Virus; *Lc* gene: an *R*-type (basic helix–loop–helix) maize transcription regulator of the anthocyanin biosynthetic pathway; rbcS 3′: the 3′-untranslated region of the Rubisco small subunit gene; restriction sites: *BglI*I, *EcoRI, Eco*RV, *Hin*dIII, *NcoI*, and *XbaI*. *Eco*RI (1), *Hin*dIII (67), *Hin*dIII (4,835), and *BglI*I sites are located outside the transposon; (**b**) somatic excision sectors of anthocyanin pigmentation observed in the aleurone layer in an *r–r* or *r–g* genetic background.

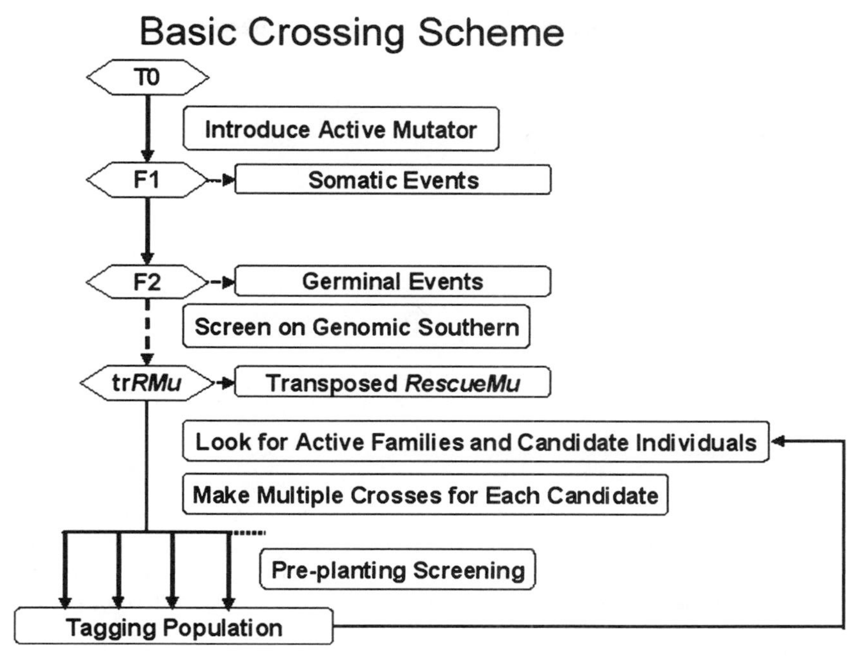

Fig. 2. Basic crossing scheme of the *RescueMu* transgenic lines. T_0: primary transgenic lines; F_1: first generation after crossing with active *Mutator;* somatic events can be scored at this generation; F_2: second generation after crossing with active Mutator; germinal events can be scored for the first time at this generation; tr*RMu*: screening for transposed *RescueMu* events that have segregated away from the original transgene loci; Tagging population: plants grown from the seed from either multiple outcrosses of a selected individual (once as ear parent and many times as pollen parent) or multiple crosses among siblings in an actively transposing tr*RMu* family.

1–2 copies of *RescueMu* without the original transgene array(s) were identified. These *RescueMu* elements that have transposed to unlinked location and segregated away from the original loci were named transposed *RescueMu* or tr*RMu*. Lineages derived from one superior tr*RMu* line have produced many generations of materials with much improved frequencies of *RescueMu* germinal insertion (up to 50–140% per plant, **Fig. 3**). Each season, large-scale DNA blot hybridization assays were conducted to identify families with high transposition rates; specific individuals with certified new transpositions were employed as pollen parents onto an anthocyanin tester line (non-Mutator) to produce tagging population for random mutagenesis of maize genes. The candidate individuals were also crossed with a non-Mutator source to generate an ear progeny. To ensure that the Mutator activity stays on the progeny and more new insertions are generated, kernels from each progeny ear were sampled and screened to find the best families prior to planting. This whole process was repeated over 4 years to generate the 45,000 plants ultimately used in the random mutagenesis and gene cloning *(2)*. In the optimal design, tagging grids of 2,304 individuals (48 rows and 48 columns) were grown form the progeny of a single *RescueMu*

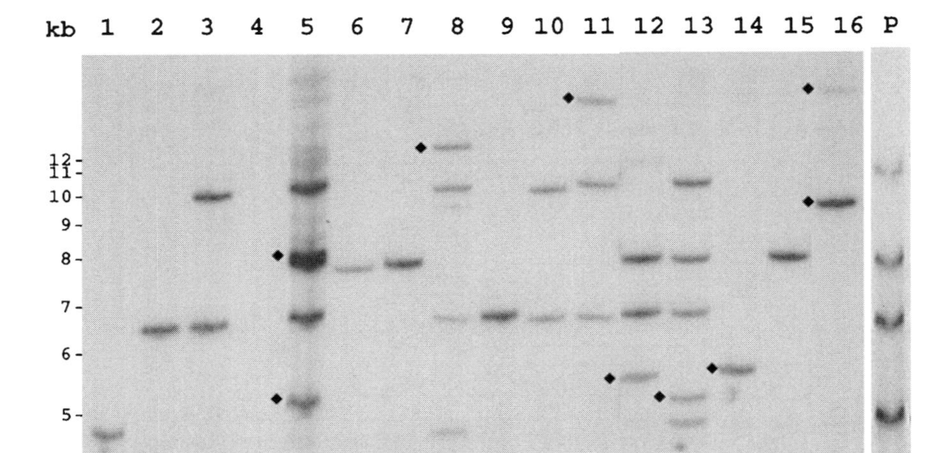

Fig. 3. DNA gel blot of 16 progeny plants (lanes 1–16) with the same *RescueMu* parent (lane P). Genomic DNA was digested with *Xba*I and hybridized with the 0.7-kb Amp gene probe. There were four parental *RescueMu* loci, which were segregated as expected in the progeny. Novel bands in the progeny, indicating new insertion events are marked (*filled diamond*). Molecular weight markers are indicated in kilobases (kb) along the *left* side of the photograph.

parent. To obtain that many individuals at least ~2,500 sibling plants were grown. As these plants had the same *RescueMu* parent, the population was segregating for the existing *RescueMu* insertion sites of that parent; typically two or three such parental insertion sites existed. The screen for tr*RMu*, thus required finding additional new insertion sites in addition to the expected parental sites.

2. Materials

2.1. Genomic DNA Extraction and Digestion

1. Urea buffer: 7 M urea, 350 mM NaCl, 50 mM Tris–HCl (pH 8.0), 20 mM EDTA (pH 8.0), 1% (w/v) *N*-lauroyl sarcosine.
2. Buffer-saturated phenol:chloroform:isoamyl alcohol (25:24:1, v/v/v) (Invitrogen, Carlsbad, CA).
3. 3 M sodium acetate (pH 4.8).
4. Restriction enzymes, RNase T1 or RNase A.
5. 13-mL polypropylene tubes (# 60.540.500) with screw caps (Sarstedt, Newton, NJ).

2.2. Agarose Gel Electrophoresis and Blotting

1. Molecular Biology Grade Low-EEO agarose (Fisher).
2. 50× TAE buffer: 242 g Tris base, 57.1 mL glacial acetic acid, 100 mL 0.5 M EDTA (pH 8.0) per liter.

3. 10× DNA loading dye: 0.25% (w/v) bromophenol blue, 0.25% (w/v) xylene cyanol, 25% (w/v) Ficoll (type 400).
4. Ethidium bromide stock: 10 mg/mL.
5. 20 (× SSC: 3 M sodium chloride, 0.3 M sodium citrate (pH 7.0).
6. Southern blotting solution: 0.4 M NaOH.
7. 3MM chromatography filter paper (Whatman, Maidstone, UK).
8. Hybond+N™ positively charged nylon membrane (GE Healthcare, Piscataway, NJ).

2.3. Hybridization and Signal Detection

1. MinElute PCR Purification Kit or the MinElute Gel Extraction Kit (Qiagen, Valencia, CA).
2. AlkPhos Direct Labeling and Detection System (GE Healthcare, Piscataway, NJ).
3. CDP-Star™ detection reagent (GE Healthcare, Piscataway, NJ or New England Biolabs, Beverly, MA).
4. BioMax Light X-ray film (Eastman Kodak, Rochester, NY) or equivalent.

3. Methods

3.1. DNA Extraction and Digestion

3.1.1. Mini-Urea Extraction of Total Genomic DNA from Maize Leaf Materials

1. Pour liquid nitrogen over a leaf piece (ca. 10 cm²). Alternatively, more tissue can be ground and stored below –80°C in a 13-mL tube and dispensed as required.
2. Use a sturdy glass rod (–1.0 cm in diameter) to break tissue into small pieces and immediately after most liquid nitrogen has evaporated, pulverize the tissue on a vortex with the glass rod in the tube for 1–2 min (*see* **Note 1**).
3. Close the cap and store the tube in a liquid nitrogen tank or in a –80°C freezer until ready for DNA extraction.
4. Transfer 0.5–1 mL packed volume of ground tissue to 2-mL microcentrifuge tube. Add 1 mL Urea Buffer, close the cap, and mix thoroughly on a vortex. Keep the tube on ice until all the samples are ready for the next step.
5. In a fume hood, add 0.7 mL phenol:chloroform:isoamyl alcohol mix to each sample (*see* **Note 2**), close the cap and invert the tubes vigorously for 1–2 min.
6. Centrifuge 5 min at 10,000 × *g* at room temperature.

7. In a fume hood, carefully transfer the upper aqueous phase with a P-1,000 pipette to a new 2-mL tube containing 100 μL 3 M sodium acetate and mix gently by pipetting.
8. Add 1 mL isopropanol and let it sit on ice without mixing for at least 5 min.
9. Gently inverting the tubes for ~1 min or until the strands of DNA become visible.
10. Spin at 3,000 × *g* for 1 min at room temperature and carefully pour off the supernatant.
11. Add 500 μL 70% ethanol and invert the tube several times.
12. Centrifuge at 10,000 × *g* for 1 min and carefully pour off the ethanol.
13. Repeat **steps 11** and **12**.
14. Centrifuge briefly (pulse).
15. Remove remaining ethanol carefully with a P-200 pipette.
16. Leave the cap open for 30 min in a laminar flow hood with the opening toward the air flow to dry the pellet.
17. Add 250 μL 10 mM Tris–HCl, pH 8.0 buffer to the dried pellet.
18. The pellet usually dissolves nicely overnight at 4°C. Alternatively, allow the pellet to rehydrate for 10 min before gently mixing by tapping the tube.
19. Store the DNA extract at –20°C or below.

3.1.2. Restriction Digestion

1. Mix 5–10 μg genomic DNA (ca. 50–75 μL) in 1× restriction buffer containing 3–5 units of each restriction enzyme per μg DNA, 500 units RNase T1 or 200 μg RNase A, and water to a final volume of 400 μL. Incubate overnight at the optimal digestion temperature.
2. Add 40 μL 3 M sodium acetate, mix, add 1 mL 100% ethanol, and mix again.
3. Incubate at 4°C or below for at least 10 min before centrifugation at 10,000 × *g* for 20 min at 4°C.
4. Decant the ethanol supernatant.
5. Add 500 μL 70% ethanol and mix by tapping or inverting the tube.
6. Centrifuge at 12,000 × *g* for 1 min at room temperature.
7. Carefully decant the ethanol supernatant.
8. Repeat **steps 5–7** one more time.
9. Follow **steps 14–16** in **Subheading 3.1.1** to dry the pellet.
10. Resuspend in 10–15 μ (L 1 × DNA loading dye (*see* **Note 3**).

3.2. Agarose Gel Electrophoresis and Southern Blotting

3.2.1. Gel Electrophoresis and UV Nicking

1. Cast a 0.8% agarose gel in 1× TAE buffer (*see* **Notes 3 and 4**).
2. Run at 20–30 V overnight (*see* **Note 5**).
3. Immerse the gel in 0.4 mg/L ethidium bromide solution with shaking for 30 min.
4. Take a picture of the gel along with a fluorescent ruler on a UV transilluminator so that the size of the band detected later on the hybridized blot can be estimated (*see* **Note 6**).
5. Perform UV nicking on a UV transilluminator for 2–3 min.

3.2.2. Alkaline Blotting

1. In a glass tray, set up a solid smooth support (e.g., an inverted gel tray) at the center of the tray and level it.
2. Prepare 2–3 sheets of 3MM filter paper, cut to the exact size of the gel, and another 2–3 bigger sheets, cut to the exact width but about 2-inches longer than the gel.
3. Pour some blotting solution over the support and lay down one sheet of the longer 3MM paper.
4. Use a clean 10-mL pipette to gently roll around on top to get rid of any bubbles in between the layers.
5. Repeat **steps 3** and **4** to overlay the rest of the longer 3MM paper.
6. Pour more blotting solution on top of the stack.
7. Carefully invert the gel up side down and overlay on top of the stack.
8. Repeat **step 4**.
9. Prepare a sheet of Hybond^{TM-} N^{+} nylon membrane, cut to the exact size of the gel and write down the date with a pencil at a corner of the membrane (*see* **Note** 7).
10. Pre-wet the membrane in water and overlay on top with the pencil marking side down so the blot with the pencil marking will be the same orientation as the gel.
11. Repeat **step 4**.
12. Repeat **steps 3–5** with 2–3 sheets of gel-size 3MM paper.
13. Top it with a full-stack of single-fold paper towels and some weight (ca. 300–500 g) on the very top (*see* **Notes 8** and **9**).
14. After blotting overnight, disassemble the stack and rinse the membrane 2–3 times with excess 2× SSC on a shaker for 10 min each.
15. DNA immobilization: Dry the membrane between sheets of 3MM paper for 30–60 min in a 65°C oven or perform UV crosslinking at the optimum energy of 120–125 mJ/cm^2.
16. Membrane can be hybridized directly or stored dried between sheets of 3MM paper.

3.3. Hybridization and Signal Detection

3.3.1. Prehybridization

1. Equilibrate membrane in 50–100 mL 2× SSC for 5–10 min at 65°C in a rotary hybridization oven while preparing hybridization solution as directed by the manufacturer (*see* **Notes 10** and **11**).
2. Drain 2× SSC and replace with hybridization solution.
3. Incubate for at least 15 min at 65°C in a hybridization oven (*see* **Note 12**).

3.3.2. Nonradioactive Probe Labeling

1. Make ample amount of probe: For detection of *RescueMu* a 0.7-kb probe for the ampicillin-resistance gene in the pBluescript plasmid was amplified under standard PCR conditions using the following primer pair: forward: 5′-CATTTTGCCTTCCTGTTTTTG-3′ and reverse: 5′-ACTCCCCGTCGTGTAGATAA-3′.
2. Purify the PCR product using the MinElute PCR Purification Kit or the MinElute Gel Extraction Kit.
3. Label 1–2 ng of purified fragment per cm^2 of membrane at 37°C for 30 min – 2 h using the AlkPhos Direct kit following the manufacturer's instructions.
4. Add labeled probe to the hybridization solution (*see* **Note 13**) and hybridize overnight at 65°C in the hybridization oven.

3.3.3. Membrane Washes and Signal Detection (see Note 14)

1. Rinse the membrane twice in 500 mL primary wash (see manufacturer's instruction) for 10–15 min each in a 65°C water bath with constant shaking.
2. Rinse the membrane twice in 500 mL 1× secondary wash (see manufacturer's instruction) for 5 min each at room temperature on a shaker.
3. Pipette CDP-Star™ detection reagent (5–10 μL/cm^2 of membrane of the ready-to-use or a freshly prepared 125 μM solution) into a clean glass tray.
4. Drain excess wash buffer from the membrane and overlay with the DNA side directly onto the detection reagent.
5. Incubate for 3 min at room temperature.
6. Transfer the membrane into a clean, smooth, transparent sealable bag (*see* **Note 15**).
7. Squeeze out excess reagent and bubbles before heat seal the bag (*see* **Note 16**).
8. Expose to a sheet of X-ray film with the DNA side of the membrane facing the film. Exposure time ranges from 10 min–3 h, depending on the signal intensity (*see* **Note 17** for second-day exposure).
9. For reprobing, the hybridized membrane can be stripped in 0.5% SDS at 65°C for 1–2 h.
10. The stripped membrane can be hybridized directly or stored dried between sheets of 3MM paper.

4. Notes

1. Avoid thawing the frozen tissue/powder before adding the extraction buffer.
2. Before adding the phenol:chloroform:isoamyl alcohol mixture, if there are any traces of tissue on the rim of the tube, wipe it clean with a sheet of Kimwipe to prevent organic solvent leakage in the subsequent steps.
3. DNA Loading Dye (Type II) and gel buffer (1× TAE) were chosen for their superior banding resolution and separation over other types of loading dyes (types I, III, and IV) and buffer *(4)*.
4. Level the gel casting and electrophoresis apparatus prior to use.
5. Depending on the type of apparatus, it might be necessary to weigh down the gel with a glass plate to prevent gel flotation. This can occur during the prolonged electrophoresis process, because bubbles may accumulate underneath the gel.
6. Picture of a successful gel, indicating good separation, digestion, genomic DNA qualities, usually has several bright, distinctive ethidium bromide-stained bands, representing the repetitive sequences, e.g., rDNA.
7. Handle the membrane with blunt-end forceps at all times.
8. The size of the paper towel should be as close as possible to the size of the gel. In cases when it is longer or wider, strips of parafilm can be placed along the side of the gel to prevent direct contact of paper towels with the blotting solution in the tray.
9. An unwanted hard-copy catalog or book is ideal for weighing down the blotting stack.
10. Heating and mixing are required to dissolve the blocking reagent. Here are some pointers for making hybridization solution on a hot plate/stirrer: (1) First, put blocking reagent and sodium chloride along with a stir bar in a flask; (2) start stirring before slowly adding in the hybridization buffer; (3) cover the flask with aluminum foil and then turn the heat to maximum; (4) in about 1 min, turn the heater off when the plate has just become scorching hot; and (5) continue stirring until completely dissolved (in ~5–10 min).
11. When there is some overlap of the membrane in the hybridization tube, make sure the direction of the membrane spiral (from the inside edge toward the outside edge) is the same as that of the tube rotation. This will avoid further curling.
12. Hybridizing at a lower temperature, e.g., at 55°C as recommended by the manufacturer, will result in a much higher background.

13. Take out a portion of the hybridization solution and mix in the labeled probe before pouring the mixture back into the hybridization tube.
14. To reduce signal background, it is very important to clean everything (container, tray, and sealable bag needed in this section) with 1% SDS followed by several rinses of water each time before use.
15. Minimize membrane handling as much as possible, especially after applying the CDP-Star™ reagent. The sealable bag can be cut open with a clean razor blade before putting the blot in.
16. Some DNA ladders may react strongly with certain probes and cause signal bleeding. To minimize bleeding, the DNA ladder should be loaded in the first or the last well and as an extra precaution, leave the lane next to the DNA ladder empty. After sandwiching the blot in the sealable bag, seal all sides except the side closest to the ladder, so excess liquid can be squeezed out through this opening.
17. When necessary, second-day exposures can be performed on the membrane after a quick rinse in the secondary wash followed by reapplication of fresh CDP-Star™ reagent.

Acknowledgments

The authors thank Dr. Manish Raizada who generated the original *RescueMu* transgenic lines used in this study. This project was supported by National Science Foundation Grant 98-72657 to V.W.

References

1. Bennetzen, J.L., Springer, P.S., Cresse, A.D., and Hendrickx, M. (1993). Specificity and regulation of the mutator transposable element system in maize. *Crit. Rev. Plant Sci.* 12, 57–95.
2. Fernandes, J., Dong, Q., Schneider, B., Morrow, D.J., Nan, G.-L., Brendel, V., and Walbot, V. (2004). Genome-wide mutagenesis of *Zea mays* L. using *RescueMu* transposons. *Genome Biol.* 5, R82.
3. Raizada, M.N., Nan, G.-L., and Walbot, V. (2001). Somatic and germinal mobility of the *RescueMu* transposon in transgenic maize. *Plant Cell* 13, 1587–1608.
4. Sambrook, J., Fritsch, E.F., and Maniatis, T. (1989) Molecular Cloning: A Laboratory Manual – 2nd edition, Chapter 6 (Nolan, C., ed.) Cold Spring Harbor, New York, NY.

Chapter 10

Tissue-Print Immunodetection of Transgene Products in Endosperm for High-Throughput Screening of Seeds

M. Paul Scott

Summary

This method allows high-throughput qualitative screening to identify seeds containing a transgene product in endosperm tissue. It is particularly useful for determining genetic segregation ratios or identifying seeds to be advanced in a breeding program. Tissue printing is used to avoid time-consuming extraction steps. Antibody-based detection of the transgene product makes this method suitable to any transgene product for which a specific antibody is available. It is possible to screen thousands of seeds per week using this method.

Keywords: Seed, Endosperm, Antibody, Protein

1. Introduction

The primary function of seed endosperm tissue is to accumulate reserves for the germinating seedling. This storage function makes endosperm an attractive target for accumulation of transgene products, because foreign proteins that accumulate in a storage tissue are less likely to interfere with development than those accumulating in other tissues. Endosperm-expressed transgenes have been used to improve grain quality *(1–4)* and as a vehicle to produce high-value proteins such as vaccines *(5)* and other high-value proteins *(6–10)*. The seed storage protein promoters are among the strongest in the plant and several have been shown to function well in transgenes.

Transgenes often do not confer a visible phenotype to the plant, but breeding and genetic experiments often require selection of transgenic plants from segregating populations.

M. Paul Scott (ed.), *Methods in Molecular Biology: Transgenic Maize, vol. 526*

DOI: 10.1007/978-1-59745-494-0_10

Therefore, laboratory methods for selection of transgenic plants are required. Selection at the seed stage is advantageous because plants that do not express the transgene are identified before planting and therefore do not need to be grown, minimizing the number of plants that need to be grown. Evaluation of seeds is also convenient because seeds can be stored many years and analyzed at a time that is convenient for the researcher.

Immunological detection of the transgene product can be used to identify transgenic seeds if an antibody that reacts with the transgene product is available. Immunological methods are attractive because general protocols *(11)* such as western blotting or ELISA can be applied to different transgene products using different antibodies specific for each transgene product of interest. This minimizes the number of protocols required when evaluating transgenes with a variety of different products.

This chapter describes a high-throughput method for identifying transgenic seeds containing endosperm-expressed transgene products. This method is not a quantitative measure of transgene levels, but it is useful for determining segregation ratios and selecting transgene-positive seeds for planting. While the pericarp is damaged by this method, we plant analyzed seeds without treatment and routinely get >80 germination. This method meets the high-throughput needs of breeding and genetics programs to identify seeds carrying a functional transgene.

2. Materials

1. Sand Paper, 150 grit.
2. Modeling clay.
3. Seed extraction buffer: 65.5 mM Tris–HCL, 3.33% SDS, 5% 2-mercaptoethanol, pH 6.8 (*see* **Note 1**).
4. Nylon-backed nitrocellulose membrane.
5. Blocking solution: 5% (w/v) skim milk powder in 1× PBS (137 mM NaCl, 2.7 mM KCl, 1.4 mM NaH_2PO_4, 4.3 mM Na_2HPO_4, pH 7.4).
6. Transgene product-specific antibody (*see* **Note 2**).
7. Antibody detection reagents: Many systems are available for this. We use an alkaline phosphatase labeled secondary antibody and a colorimetric detection kit (both from Bio-Rad, Hercules, CA) according to the manufacturer's directions.

3. Methods

Screening seeds by sampling the endosperm is desirable because it yields a small amount of tissue for analysis and the seed remains viable for planting. Removing the tissue and extracting the compounds of interest can be time consuming because it involves collecting and weighing tissue, extraction, and centrifugation. Tissue-printing involves first exposing the tissue of interest. We do this by abrading the seed with sand paper. The exposed surface is then moistened with the extraction buffer and pressed onto blotting media. Enough of the transgene product is transferred to the paper to be detected with immunological methods. Because it is not necessary to collect and extract the tissue, tissue-printing is much faster than traditional extraction methods. We routinely process 500 seeds in a batch.

Positive and negative controls should be included on each blot. High background signal and levels of the transgene product near the limit of detection of the method can make distinguishing positive and negative tests difficult. Background signal can sometimes be reduced using the optional protocol for preadsorption of the antibody with nontransgenic corn endosperm.

3.1. Tissue-Print Blotting

1. Prepare the seeds by numbering them with a pencil.
2. Cut the nylon-backed nitrocellulose membrane to the size of the tray you will be processing the blot in. Draw a 1-cm grid on the paper with a soft pencil. Wear gloves when handling the membrane.
3. Make a long (about 2 cm/seed), 5–7-mm diameter cylinder of modeling clay. Grasp a seed firmly between thumb and forefinger with the tip cap oriented toward the hand so that the crown of the seed is exposed. Abrade the seed by rubbing it on a piece of sand paper placed on the laboratory bench so that about 1 mm of tissue is removed to create a flat top on the seed. Use a different place on the sand paper for each seed to avoid contamination between seeds.
4. Press the seed into the modeling clay with the abraded surface up (**Fig. 1**).
5. Add 3–5 ml (depending on seed size) of seed extraction buffer to the sanded surface of the seed.
6. Wait for 2–3 min and add seed extraction buffer as before.
7. Wait approximately 1–2 min and then press the abraded surface of each seed firmly into one of the squares marked on the nitrocellulose membrane (*see* **Note 3**).
8. Allow the membrane to dry. It can be stored indefinitely before processing.

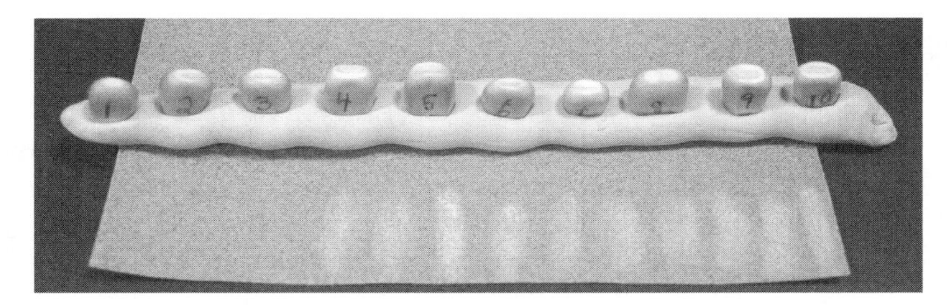

Fig. 1. Abraded kernels in modeling clay ready for addition of seed extraction buffer. The clay is resting on the sandpaper used to abrade the kernels. Marks on the sandpaper are from abrading the kernels.

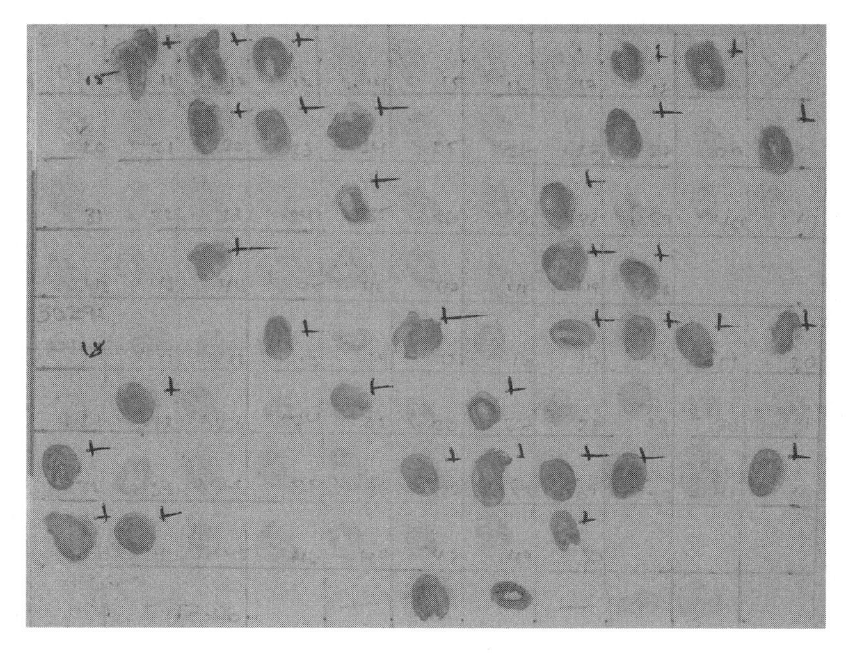

Fig. 2. A typical blot developed with an alkaline phosphatase-conjugated secondary antibody and colorimetric detection system. A plus sign indicates seeds that are considered to contain the transgene product. The bottom row was blotted with two positive control kernels (*bottom center*) and two negative control kernels *(bottom right*).

9. Process the membrane by placing it in a shallow tray containing blocking solution so that the membrane is just covered by the solution. Rock gently for 1 h.
10. Add the antibody to the solution covering the membrane. If using a preadsorbed antibody (*see* **Subheading 3.2**), replace the blocking solution with the preadsorbed antibody. Otherwise, the antibody can be added directly to the blocking solution (*see* **Note 4**). Allow the antibody to bind with gentle rocking for 5 h.
11. Develop the blot using your choice of antibody detection reagents according to the manufacturer's protocol. We use

an alkaline phosphase-labeled goat secondary antibody (Bio-Rad, Hercules, Ca) and a colorimetric alkaline phosphatase detection kit (Bio-Rad, Hercules, Ca). A typical blot is illustrated in **Fig. 2**.

3.2. Preadsorption to Remove Nonspecific Antibodies (Optional)

1. Soak nontransgenic kernels in water for 5 min and manually peel off the pericarp. Remove the embryos using a razor blade or scalpel. Grind the remaining tissue (mostly endosperm) finely in a coffee grinder or mortar and pestle.
2. Extract this ground tissue at 100 mg tissue/ml of seed extraction buffer for 20 min to 1 h. The total volume should be sufficient to just cover the blots that you intend to process with this antibody. Centrifuge and remove supernatant.
3. Add the antibody to this solution and incubate for 5 h. The resulting solution can be added to the blocked membrane.

4. Notes

1. 2-Mercaptoethanol is volatile and toxic. Add it to the solution immediately prior to use. Wear gloves and work in a fume hood when using this solution.
2. It is critical that the antibody is specific to the transgene because it is not possible to distinguish nonspecific interactions from specific ones in this method. The antibody used should react only with the transgene product in a western blot of transgenic kernels. If nonspecific interactions are detected, they can sometimes be removed by preadsorbing the antibody with nontransgenic endosperm as described in the optional method in **Subheading 3.2**.
3. A convenient way to make the tissue prints is to cut the modeling clay between each seed and pick each seed up by the clay. The goal is for the seed to be wet when it is blotted, but not so wet that the solution spreads on the blotting membrane. Different seeds soak up solution at different rates and the volume of seed extraction buffer and time to printing can be adjusted to account for this. If a seed does not leave a moist spot on the paper it is probably too dry to give a good signal. It can be remoistened with seed extraction buffer and blotted again.
4. The concentration of antibody should be optimized. Generally, a concentration that works well for a western blot will work well for this protocol. This procedure works with mono- or polyclonal antibodies.

Acknowledgments

The author wishes to thank Mrs. Merinda Struthers for technical assistance.

References

1. Lai, J. S. and Messing, J. (2002) Increasing maize seed methionine by mrna stability. *Plant J.* 30 395–402.
2. Yang, S. H., Moran, D. L., Jia, H. W., Bicar, E. H., Lee, M. and Scott, M. P. (2002) Expression of a synthetic porcine alpha-lactalbumin gene in the kernels of transgenic maize. *Transgenic res.* 11 11–20.
3. Huang, S., Kruger, D. E., Frizzi, A., D'Ordine, R. L., Florida, C. A., Adams, W. R., Brown, W. E. and Luethy, M. H. (2005) High-lysine corn produced by the combination of enhanced lysine biosynthesis and reduced zein accumulation. *Plant Biotechnol. J.* 3 555–569.
4. Yu, J., Peng, P., Zhang, X., Zhao, Q., Zhy, D., Sun, X., Liu, J. and Ao, G. (2004) Seed-specific expression of a lysine rich protein < em > sb401 < /em > gene significantly increases both lysine and total protein content in maize seeds. *Mol. Breed.* 14 1–7.
5. Streatfield, S. J., Jilka, J. M., Hood, E. E., Turner, D. D., Bailey, M. R., Mayor, J. M., Woodard, S. L., Beifuss, K. K., Horn, M. E., Delaney, D. E., Tizard, I. R. and Howard, J. A. (2001) Plant-based vaccines: Unique advantages. *Vaccine* 19 2742–2748.
6. Lamphear, B. J., Barker, D. K., Brooks, C. A., Delaney, D. E., Lane, J. R., Beifuss, K., Love, R., Thompson, K., Mayor, J., Clough, R., Harkey, R., Poage, M., Drees, C., Horn, M. E., Streatfield, S. J., Nikolov, Z., Woodard, S. L., Hood, E. E., Jilka, J. M. and Howard, J. A. (2005) Expression of the sweet protein brazzein in maize for production of a new commercial sweetener. *Plant Biotechnol. J.* 3 103–114.
7. Woodard, S. L., Mayor, J. M., Bailey, M. R., Barker, D. K., Love, R. T., Lane, J. R., Delaney, D. E., McComas-Wagner, J. M., Mallubhotla, H. D., Hood, E. E., Dangott, L. J., Tichy, S. E. and Howard, J. A. (2003) Maize (*Zea mays*)-derived bovine trypsin: Characterization of the first large-scale, commercial protein product from transgenic plants. *Biotechnol Appl. Biochem.* 38 123–130.
8. Zhong, G. Y., Peterson, D., Delaney, D. E., Bailey, M., Witcher, D. R., Register Iii, J. C., Bond, D., Li, C. P., Marshall, L., Kulisek, E., Ritland, D., Meyer, T., Hood, E. E. and Howard, J. A. (1999) Commercial production of aprotinin in transgenic maize seeds. *Mol. Breed.* 5 345–356.
9. Witcher, D. R., Hood, E. E., Peterson, D., Bailey, M., Bond, D., Kusnadi, A., Evangelista, R., Nikolov, Z., Wooge, C., Mehigh, R., Kappe, W., Register, J. and Howard, J. A. (1998) Commercial production of beta-glucuronidase (gus): A model system for the production of proteins in plants. *Mol. Breed.* 4 301–312.
10. Hood, E. E., Witcher, D. R., Maddock, S., Meyer, T., Baszczynski, C., Bailey, M., Flynn, P., Register, J., Marshall, L. and Bond, D. (1997) Commercial production of avidin from transgenic maize: Characterization of transformant, production, processing, extraction and purification. *Mol breed* 3 291–306.
11. Harlow, E. and Lane, D. (1999) *Using Antibodies: A laboratory Manual.* Cold Spring Harbor Laboratory Press, Cold Spring Harbor, New York.

Chapter 11

Determination of Transgene Copy Number by Real-Time Quantitative PCR

Colin T. Shepherd, Adrienne N. Moran Lauter, and M. Paul Scott

Summary

Efficient methods to characterize transgenic plants are important to quickly understand the state of the transformant. Determining transgene copy number is an important step in transformant characterization and can differentiate between complex and simple transformation events. This knowledge can be extremely useful when determining what future experiments and uses the transgenic lines can be utilized for. The method described here uses real-time quantitative PCR to determine the transgene copy number present in the genome of the transformant. Specifically, this method measures the relative transgene copy number by comparing it with an endogenous gene with a known copy number. This method is a quick alternative to the Southern blot, a method that is commonly used to determine gene copy number, and is effective when screening large numbers of transformants.

Keywords: Transgenic plants, Quantitative PCR, Gene copy number

1. Introduction

Transgenic plants are created for a number of reasons that include both basic research and industrial commercialization applications. The number of times the transgene inserts into the genome (copy number) directly relates to the gene's expression level, insert stability, and inheritance. If commercialization of the end product is the objective, the transgenic plants need to be characterized correctly because complex events that contain multiple transgenes are more difficult to commercialize. Complex transformation events that contain truncated inserts and inverted repeats commonly occur *(1)*, as opposed to simple insertions containing one copy of the transgene. Determination of transgene copy number to distinguish between complex and simple events, zygosity,

M. Paul Scott (ed.), *Methods in Molecular Biology: Transgenic Maize, vol. 526*

DOI: 10.1007/978-1-59745-494-0_11

and correct characterization of the transgene can save time and resources in future generations.

Southern blotting has traditionally been used to characterize genomic DNA and determine gene copy number. In this method, genomic DNA is digested with restriction enzymes and hybridized with a DNA probe. The resulting band pattern provides information about the number of transgene copies. However, this method is time consuming and is difficult to perform on a large scale, which makes it inefficient to use when analyzing large numbers of transgenic plants. In addition, complex events that contain multiple transgenes in concatamers or other complex arrangements may result in Southern blots that are difficult to interpret. An alternative method that is both faster and easier to perform on large numbers of plants is real-time quantitative PCR (RT-qPCR). Where a typical Southern blot protocol takes days to perform after the DNA has been extracted, RT-qPCR reactions take hours to run and yield data in digital form for analysis. However, there may be cases where RT-qPCR is not feasible, or information about the arrangement of the transgenes in the genome is required, and in these cases nonradioactive Southern blot protocols can be used to compliment RT-qPCR (*see* the chapter "Nonradioactive Genomic DNA Blots for Detection of Low Abundant Sequences in Transgenic Maize").

PCR amplifies DNA in an exponential fashion. The amount of product is eventually limited in each reaction by reagents such as primers and dNTPs, resulting in a plateau phase in which the amount of product produced is limited. Each reaction then produces different amounts of product, independent of starting template amounts. Thus, the amount of PCR product is related to the template copy number during the logarithmic phase of amplification. It is only during this phase of amplification that it is possible to extrapolate back to the amount of starting template.

In real-time PCR, detection of product is observed during thermal cycling by monitoring fluorescence in the PCR reaction. There are two types of detection chemistries available for real-time PCR, gene-specific fluorescent-labeled probes, and nonspecific dsDNA binding dyes. The gene-specific probes include TaqMan® probes or molecular beacons and are a more sensitive method to quantify gene copy number. The use of gene-specific labeled probes requires purchase of the fluorescent probe in addition to the primers, thereby increasing the cost of the experiment. Rather than utilizing a gene-specific probe, we employed a fluorescent nonspecific dsDNA binding dye (SYBR green), which is incorporated into newly amplified DNA. Both specific and nonspecific PCR products will bind SYBR green, but a simple dissociation curve analysis at the end of the PCR run will determine whether nonspecific products are present (*see* **Note 1**).

A threshold cycle (C_t) for each amplification is assigned by the RT-qPCR software according to the cycle in which the fluorescence signal is significantly greater than the background noise. Samples with more starting template (i.e., higher copy number) will generate enough amplicons to cross the threshold level first, and therefore have a lower C_t value. The C_t value is therefore related to the copy number of the template DNA.

The PCR efficiency in part determines the number of cycles required to amplify a given target. PCR efficiency corresponds to the proportion of template molecules that are doubled every cycle. Efficiency is determined using a standard curve consisting of a dilution series for each primer pair used. The standard curves for two templates to be compared should be parallel with an R^2 close to 1.

Although it is possible to derive the starting amount of template present in a PCR reaction knowing the amount of product produced by relating it to a standard curve (absolute quantitation), it is often easier to compare the copy number of a gene of interest to the copy number of a gene with a known copy number (relative quantitation) *(2)*. The 2^{-Ct} method can be used to calculate the relative differences between the control and experimental samples, and requires careful selection of an endogenous control gene of known copy number *(3)*. Bubner et al. *(4)* used a 2^{-Ct} approach *(3)*, which compared a calibrator line (a transgenic line shown to have one copy by Southern blot analysis) with the experimental lines, and found that distinguishing between 1- and 2-copy lines was the limit of RT-PCR. However, Ingham et al. *(5)* found the correlation between Taqman assay and Southern blots to be much better, presumably due to the higher accuracy gained by using a gene-specific probe.

The method described here in detail is the relative quantitation method that determines gene copy number by comparing the transgene amplification to that of an endogenous control gene that has a known gene copy number. This method meets the needs of transgenic plant producers as it is easy to perform and can be used to analyze a large number of transformants.

2. Materials

1. MX3000P real-time PCR system (Stratagene, La Jolla, CA).
2. Real-time PCR reagents Brilliant® SYBR® Green master mix (Stratagene, La Jolla, CA).
3. Optically clear PCR tubes 200 μL (Stratagene, La Jolla, CA).

4. DNA primers for transgene and endogenous control gene (0.5 μM final concentration).
5. Genomic DNA in tenfold dilution series for standard curve determination.

3. Methods

1. Set up two sets PCR reactions, each containing a dilution series (*see* **Note 2**) of the genomic DNA as the template. Prepare one set of reactions to amplify the gene of interest and the other to amplify the known copy number control sequence (*see* **Note 3**). Each reaction should be run in triplicate. PCR reactions that do not contain genomic DNA (no template control) or DNA polymerase should be included as negative controls.
2. The quantitative real-time PCR analyses were performed using the MX3000P real-time PCR system (Stratagene, La Jolla, CA).
3. PCR reactions contained 12 μL of Brilliant® SYBR® Green master mix (Stratagene, La Jolla, CA), 12 μL of ddH_2O, 1 μL of each primer (0.5 μM final concentration), and 1 μL of DNA (DNA amount as determined by the dilution series).
4. The cycling parameters were as follows: 95°C for 10 min, 40 cycles of 95°C for 30 s, 55°C for 60 s, and 72°C for 30 s.
5. Verify that the PCR product of each reaction is of the expected size by ethidium bromide agarose gel electrophoresis and/or has a single peak in the dissociation curve analysis. Primer-dimers and spurious amplification products will interfere with the procedure (*see* **Note 1**).
6. Calculate the PCR efficiency for the gene of interest and the copy number control gene as follows: Calculate the slope of the line fit to a plot of Ct vs. amount of template. The PCR efficiency is calculated with the following equation:

   $\text{Efficiency} = 10^{(-1/\text{slope})} - 1.$
7. PCR efficiencies should be greater than 90% (*see* **Note 4**).
8. To calculate the ratio of the copy number of the gene of interest to the reference gene, select a template concentration in which both the gene of interest and the copy number control gene amplified well and apply the following equation:

$$\text{Ratio} = \frac{1+\text{Efficiency}_{\text{gene of interest}}^{(\text{Ct gene of interest})}}{1+\text{Efficiency}_{\text{control}}^{(\text{Ct control})}}.$$

4. Notes

1. Real-time PCR products can be measured by utilizing sequence-specific fluorescent probes or by nonspecific dsDNA binding dyes. In this method, SYBR green is used as the non-specific dye, which has an excitation wavelength of 497 nm and an emission wavelength of 520 nm. Because the SYBR green is nonspecific, it will also bind to primer-dimers, so it is necessary to optimize the PCR reaction to reduce or eliminate the primer-dimers that will interfere with the reaction. A dissociation curve analysis (or melting curve analysis) can be performed to determine whether primer-dimers exist in your reaction. Multiple peaks in the melting curve indicate the presence of nonspecific products and/or primer-dimers, and reaction conditions should be optimized to minimize this.
2. Prepare a twofold serial dilution series with at least five points. We suggest a range between 1 and 100 ng of genomic DNA.
3. The endogenous control gene should be a gene that is known to be present in a single copy. We have used globulin-1 of maize as our endogenous control gene [*(6)*, GenBank accession M24845). The *glb-1* control gene primer sequences are forward 5′-CACTGTGGAACACGACAAAGTCTG-3′ and reverse 5′-CTCACCATGCTGTAGTGTCACTGTGAT-3′. Other examples of single copy genes are *Adh*1 (accession number X04050), *hmg*a (accession number AJ131373), and *iVr*1 (accession number U16123) *(7)*.
4. If the PCR efficiency of the copy number control gene is less than 90%, then optimize the PCR reaction by changing reaction conditions such as annealing temperature, magnesium concentration, or the elongation and acquisition times. Designing new primers may be necessary if acceptable efficiency cannot be obtained by changing reaction conditions. Lastly, a different copy number control gene may be selected if a PCR efficiency of >90% is not achieved.

References

1. Smith, N., J. Kilpatrick, et al. (2001). Superfluous transgene integration in plants. *Crit. Rev. Plant Sci.* 20:215–249.
2. Winer, J, Jung, C. K., Shackel, I., and Williams, P.M. 1999. Development and validation of real-time quantitative reverse transcriptase-polymerase chain reaction for monitoring gene expression in cardiac myocytes in vitro. *Anal. Biochem.* 270:41–49.
3. Livak, K. and Schmittgen, T.D. 2001. Analysis of Relative Gene Expression Data Using Real-Time Quantitative PCR and the 2^{-CT} Method. *Methods* 25:402–408.
4. Bubner, B., Gase, K., and Baldwin, I. 2004. Two-fold differences are the detection limit for determining transgene copy numbers in plants by real-time PCR. *BMC Biotechnol.* 4:14.
5. Ingham, D. J., Beer, S., Money, S. and Hanson, G. (2001) Quantitative real-time PCR assay for determining transgene copy number in transformed plants. *Biotechniques* 31: 132–140.

6. Belanger, F.C. and Kriz, A.L. 1989. Molecular characterization of the major maize embryo globulin encoded by the *Glb1* gene1. *Plant Physiol.* 91:636–643.
7. Hernández, M., Duplan, M., Berthier, G., Vaïtilingom, M., Hauser, W., Freyer, R., Pla, M., and Bertheau,Y. 2004. Development and comparison of four real-time polymerase chain reaction systems for specific detection and quantification of *Zea mays* L. J. Agric. *Food Chem.* 52: 4632–4637.

Part V

Breeding with Transgenes

Chapter 12

Herbicide Resistance Screening Assay

Joan M. Peterson

Summary

Herbicide resistance screening is a method that can be used not only to determine presence of the enzyme, phosphinothricin acetyltransferase, encoded by either the *Bar* or the *Pat* gene in transgenic maize, but also to assess the inheritance ratio of those genes in a segregating population. Herbicide screening can also be used to study linkage of a transgene of interest that was cotransformed with the herbicide resistance marker gene.

By combining the herbicide screen assay with a PCR-based screen of leaf tissue DNA for the presence of both the *Bar* or the *Pat* gene marker and a cotransformed transgene of interest from the same seedling tissue and maintaining that seedling identity, the researcher can identify linkage or the possible breakdown in linkage of the marker gene and the transgene of interest. Further, the occurrence of "DNA silencing" can be evaluated if an individual seedling that was susceptible to the applied herbicide nonetheless gave PCR data that indicated presence of the gene responsible for herbicide resistance. Similarly, "DNA silencing" of the gene of interest may be investigated if the seeds can be screened and scored for that phenotypic trait in a nondestructive manner prior to planting.

Keywords: Herbicide resistance, Screening assay, *Bar* gene, *Pat* gene, Phosphinothricin acetyltransferase (PAT), Phosphinothricin tripeptide, Phosphinothricin, Bialaphos, Glufosinate

1. Introduction

Herbicide resistance screening is a method that can be used to determine the presence of the enzyme, phosphinothricin acetyltransferase (PAT), that confers resistance to bialaphos or phosphinothricin-based herbicides in transgenic calli and plants *(1–6)*. Herbicide resistance screening can be used to easily assess the inheritance of the *Bar* or the *Pat* gene in a segregating population and to study linkage of a transgene of interest that was cotransformed with the marker gene. Often a transgene of interest will be tightly linked to the herbicide resistance gene in

M. Paul Scott (ed.), *Methods in Molecular Biology: Transgenic Maize, vol. 526*

DOI: 10.1007/978-1-59745-494-0_12

transformation events. When this is the case, herbicide screening can be an effective method to screen for plants likely to contain a transgene of interest. If the transgene of interest is difficult and costly to detect, screening for herbicide resistant plants can save time and money. Herbicide screening is effective in both seedling and mature plants.

The enzyme, PAT, is encoded by the *Bar* (*bia*laphos *r*esistance) gene from the soil bacterium *Streptomyces hygroscopicus* and by the *Pat* (*p*hosphinothricin *a*cetyl *t*ransferase) gene from *Streptomyces viridochromogenes* *(3, 7–10)*. PAT is produced by *Streptomycete* species to prevent autotoxicity to the secondary metabolite and antibiotic, bialaphos, naturally produced by the bacteria as they enter their stationary phase of growth *(3, 11)*. Bialaphos is a tripeptide made up of two alanine residues coupled to the amino acid, phosphinothricin, an analogue of l-glutamic acid, to form the tripeptide *(10, 12, 13)*. Bialaphos becomes toxic to other microbes or plant cells when taken up and metabolized by endogenous intracellular peptidases which remove the alanine residues and release phosphinothricin. Phosphinothricin, as an analog of l-glutamic acid, is an inhibitor of glutamine synthetase (EC 6.3.1.2), a key plant enzyme in the assimilation of ammonia and regulation of nitrogen metabolism. The inhibition of glutamine synthetase causes ammonia ions to accumulate and cell death follows. In cells containing PAT, however, the free ammonium groups of phosphinothricin are acetylated and its inhibition of glutamine synthetase and toxicity are prevented. Metabolism of phosphinothricin and glufosinate is reviewed in detail by Dröge *(11)* and Hoerlein *(14)*.

By combining the herbicide screen assay with a PCR-based screen of leaf tissue DNA for the presence of both the marker gene and a cotransformed transgene of interest from tissue of the same seedling and maintaining that seedling identity, the researcher can identify linkage or the possible breakdown in linkage of the marker gene and transgene of interest. Further, the possible occurrence of "DNA silencing" can be evaluated if an individual seedling that was susceptible to application of the herbicide nonetheless gave PCR data that indicated presence of the gene responsible for herbicide resistance. Similarly, "DNA silencing" of the gene of interest may be investigated if the seeds can be screened and scored for that phenotypic trait in a nondestructive manner prior to planting.

When the transgene inserts at a single locus, it normally behaves as a typical dominant gene and displays simple Mendelian inheritance in progeny. Zygosity of the transgene in an individual can be tested by self-pollination or cross-pollination to a nontransgenic line and conducting a segregation analysis on the progeny using the herbicide screening assay and chi-squared statistical test *(15, 16)*. Chi-squared analysis produces a ratio that gives a best approximation for assigning a segregation model of 3:1 or 1:1. In order to assign an inheritance ratio to the trait that is statistically significant, the number of individuals or seeds that must be tested or screened for herbicide resistance is critical.

In addition to its use as a selection marker for maize transformation, the *Bar* gene has been used to develop herbicide-resistant crops including canola, maize, cotton, rice, and soybean that have been approved for use and are available to farmers in North America. The use of herbicide-resistant crops has been recently reviewed by Duke *(17)*.

Chemically synthesized phosphinothricin, commonly known by its synonym, glufosinate ammonium, the ammonium salt of phosphinothricin, or by its chemical name, butanoic acid, 2-amino-4-(hydroxyl-methylphosphinyl)-, monoammonium salt, is the active ingredient (a.i.) in commercial formulations marketed under the tradenames Liberty®, Ignite®, and Rely® by Bayer CropScience, and Finale® and Derringer® by Bayer Environmental (Research Triangle Park, NC) in the US, and by Bayer CropScience in some other countries as Basta®, Tepat®, and Harvest®. Phosphinothricin in the tripeptide form, bialaphos, is produced by fermentation of *S. hygroscopicus* and is available marketed as Herbiace® and Bilanafos® by Meiji Seika Kaisha Ltd (Japan) and by Duchefa Biochemie B.V. (Haarlem, The Netherlands), available through Gold Bio Technology, Inc., (St. Louis, MO) in the US. The effectiveness of four common phosphinothricin-based herbicides used in maize transformation has been investigated by Dennehey et al. *(18)* who reported bialaphos and its commercial formulation, Herbiace®, to be more effective than glufosinate or Basta®, its commercial formulation.

2. Materials

2.1. Seed Selection

1. Confirm parental genotypes with respect to herbicide resistance gene, and, if possible, include in the assay. Also, include seed from a nontransgenic seedlot (positive control) and from a known herbicide-resistant transgenic seedlot (negative control) among the set to be treated with herbicide and among the set that will not be treated. Seeds should be inspected to insure that the seeds are mature and that embryos are intact and the seeds are not otherwise cracked or damaged.
2. The number of seeds and replicates required to be tested are dependent upon the objective of inquiry (*see* **Note 1**).

2.2. Planting Materials

1. A planting mix containing soil, sand, and peat in equal proportions (1:1:1) is recommended.
2. Containers for planting can include pots or trays that will accommodate the number of plants required to be tested. Plastic trays approximately 3 in. deep × 9 in. wide × 20 in. long (with drainage holes) will hold approximately 50 seeds and are widely available at local plant nursery outlets. Sterilization of soil mix or containers is not required though containers should be clean.

2.3. Herbicide and Applicator

1. The active ingredient in the herbicide to be used must be compatible with the resistance mechanism conferred by the marker gene. Herbicides containing active ingredients in addition to glufosinate-ammonium, for example, Liberty ATZ® which contains a second active ingredient, atrazine (Bayer CropScience), should be avoided.
2. A surfactant such as Tween® 20 (ICI Americas, Inc., Wilmington, DE) is necessary to include when Liberty® is used but is not required when Basta® is used.
3. A hand-held, plastic spray bottle which can hold up to a liter of solution can serve as an applicator. Use separate spray bottles for water, water and surfactant, and herbicide treatments.
4. Always read the material safety data sheet (MSDS) for the herbicide used. Note the proper personal protective gear that should be worn when handling and spraying the herbicide and also when handling the pots and seedlings after spraying. Note proper and safe storage conditions for the unused herbicide and for the working solutions.

3. Methods

This method describes herbicide screening on seedlings produced in a greenhouse or growth chamber. Adaptations for screening mature plants are described in **Note 2**.

3.1. Preparation of Samples and Trays

1. Fill pots or trays with the 1:1:1 soil–peat–sand mix and make individual wells into the mix or evenly spaced furrows approximately 1 in. deep across the length of the container. Dampen the mix with tap water prior to planting.
2. Plant the seeds into the wells or furrows and cover with the soil mix. Place pots or trays into a controlled environment (*see* **Note 3**).

3. Herbicide control samples should be planted into separate pots or trays rather than into rows in the same container that will receive the herbicide application. In separate containers, the controls can be more easily protected from herbicide drift during application. Return all containers to the common growing area after application in order to maintain uniform growth conditions.

3.2. Culture Conditions

1. Water soil mix thoroughly after planting with tap water using sprinkling cans or a garden hose fitted with sprinkling nozzle and again only as the soil mix surface dries out. After emergence of the seedling (approximately 3–5 days) use care to apply water only to the soil surface and avoid getting water onto the seedling leaves. This is particularly important after the herbicide application.
2. For optimum growth conditions, supply seedlings with 16/8 h day/night and 28–30°C/22–24°C, respectively (*see* **Note 3**).

3.3. Herbicide Application

1. Rate(s) of application should be determined empirically, as an appropriate rate depends on the herbicide resistance construct properties as well as species, genotype, environment, and age of the seedlings *(19, 20)*. The application rate must also take into account concentration of the active ingredient associated with the individual commercial product used. Published rates vary from 250 mg/L to 500–2,000 mg phosphinothricin/L to 3,000 mg/L *(2, 4, 5)*.
2. In the case of limited seed supply, an empirical determination of rate can be carried out on as few as 3–5 seedlings per rate.
3. The following example is given for preparation of 1 L (1,000 ml) of a 2.0% solution of Liberty® (18% glufosinate, a.i.) with 0.1% Tween® 20 as a surfactant:

Mix thoroughly.

Reagent	Concentration	Formula	Volume to add
Liberty	(18% a.i.)	2.0% 0.020 × 1,000 ml	20.0 ml Liberty
Tween® 20	0.1%	0.001 × 1,000 ml	1.0 ml Tween® 20
			979.0 ml water

[18% a.i. × 20 ml Liberty = 0.18 × 20 ml = 3.6 g or 3,600 mg/L, a.i.]

4. Prepare and apply herbicide solution to seedlings at the third to fourth leaf stage, approximately 7–10 days after planting. One application should be sufficient.
5. Include an herbicide solution control treatment that contains water alone and/or water with only the surfactant, Tween® 20, 0.1%.
6. The mixture should be made at time of use and be used within a week.
7. Spray with a hand-held, plastic spray bottle making sure to coat the plants well. Control for uniform application of herbicide or control solution by spraying from all sides and noting formation of spray droplets forming at tip of leaf. Spray load/time is sufficient when droplets are observed built up at the tips of the leaves.
8. Manage overspray by hanging/draping a plastic sheet or cardboard barrier between flats or prepare a separate, protected area where flats can be sprayed individually and then be returned to the common area.
9. Herbicide handling and disposal should be in accordance with MSDS obtained from manufacturer. With respect to persistence and mobility in the soil, glufosinate has been shown to have a relatively short half-life of 7 days in soil as it is rapidly degraded by soil microbes.

3.4. Scoring Resistant/Susceptible Phenotypes

Monitor plants and symptoms closely, particularly when determining rates of application. Score the plants when death of the negative controls is obvious, generally at 7–10 days after herbicide application. Symptoms will include chlorosis and wilting usually within 3–5 days after application. The rate of symptom development is increased by optimal growth conditions including bright light, high humidity, and moist soil. Continue to evaluate plants daily to every few days over a several week period to note progress of the symptoms. An effective application rate should be the minimum concentration that also allows unambiguous interpretation of symptoms in a transgenic seedlot (*see* **Note 4**).

We have noted in some transgene events a range in severity of symptoms among progeny (**Fig. 1**). Some individuals displayed chlorotic symptoms within 48 h that led to death, some were initially resistant but failed to continue growing, and some were completely resistant and comparable to unsprayed controls. In the case of seedlings that initially appeared resistant but failed to thrive, PCR data indicated the *Bar* gene was present with band intensity proportional to that of the degree of resistance in the surviving seedlings (unpublished results). A range in response symptoms may reflect a range in herbicide resistance gene expression which could be due to differences in copy number or genome position effects among other things.

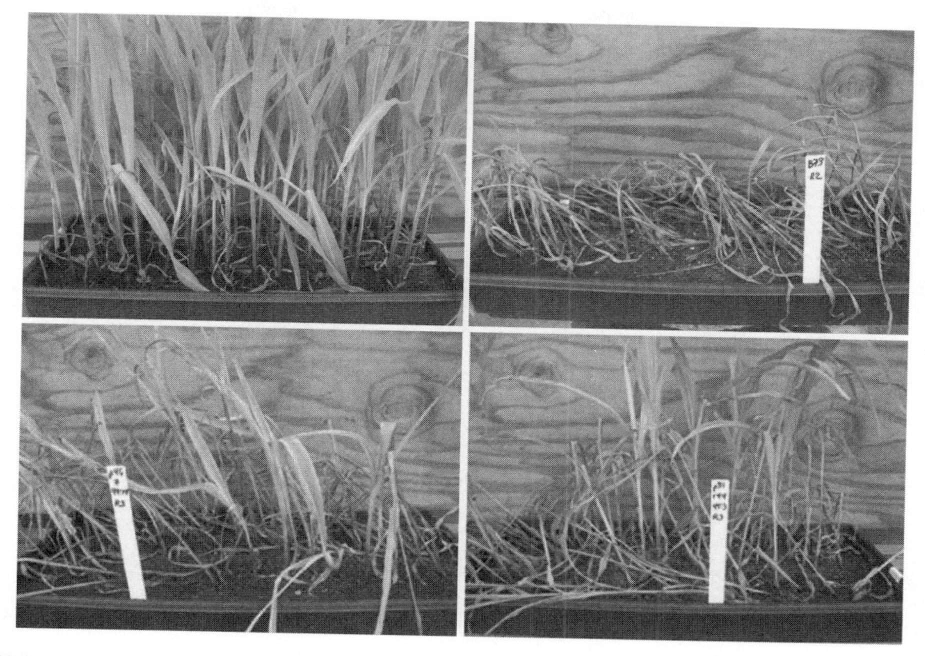

Fig. 1. Maize seedlings 7 days after application of the glufosinate ammonium herbicide, Liberty®. Nontransgenic seedlings shown in *upper left* received no herbicide treatment, nontransgenic seedlings shown in *upper right* photo received the same treatment as seedlings of transgenic seedlots shown in *lower left* and *lower right*. Seedlings resistant to the herbicide shown for the transgenic seedlots were scored positive for the *Bar* gene. All conditions were as those outlined in **Subheading 3** (*see Color Plates*).

Statistical Analysis

Use the following formula to yield the chi-squared value:

$$X^2 = \Sigma \frac{(\text{observed} - \text{expected})^2}{\text{expected}}.$$

The calculated value is then compared against a table of chi-squared values to determine a level of significance that can be associated with the calculation and sample size and acceptance of the hypothesis regarding a 1:1 or 3:1 segregation ratio.

4. Notes

1. Special consideration should be given to the number of seeds per seedlot necessary for planting if poor germination or poor seedling vigor are expected. A small subset of seeds may be planted before the herbicide testing is conducted in order to determine the germinability of the seedlot. The number

Fig. 2. Mature maize leaves from two different plants in the field treated with herbicide solution as described in **Note 2**. *Upper* leaf in *right hand* shows no symptoms of herbicide damage whereas *lower* leaf in *left hand* shows severe damage (*see Color Plates*).

of seeds subsequently planted may or may not need to be increased in order to take into account the percent of viable seeds. If there is a limited number of seeds and a need to test germination percentage, the germination test could be coupled with the rate test as a way of utilizing the seedlings for both purposes.

2. Mature plants in the field or greenhouse can be quickly screened using the following method. Mark a leaf to be screened by cutting off the end so that it is easily differentiated from the nontreated leaves. Dip the cut leaf into a wide mouth bottle containing herbicide so that about 4 in. of the leaf is exposed to herbicide. In about a week, the portion of the leaf exposed to herbicide should die (**Fig. 2**). Gloves and lab coat should be worn for this procedure.
3. Lights should be maintained an adequate distance above seedlings to provide ample light. Under greenhouse conditions, the time of year is very important. If resistant plants are to be grown on to maturity, the researcher should avoid planting in November or early December. Because herbicide symptoms develop most readily under optimum growing conditions, conversely and if necessary, symptom development can be slowed by providing less optimum conditions.

4. Troubleshooting symptoms: if the positive (transgenic) control is not resistant to the herbicide, the herbicide concentration may be too high, or the wrong herbicide or incorrect control may have been used. If the negative (nontransgenic) control shows resistance to the herbicide, the herbicide concentration may be too low, or the wrong herbicide or incorrect control may have been used. In either case, herbicide stock formulation and preparation procedure should be checked.

References

1. Spencer, T. M., Gordon-Kamm, W. J., Daines, R. J., Start, W. G., and Lemaux, P. G. (1990) Bialaphos selection of stable transformants from maize cell culture. *Theoretical and Applied Genetics* 79, 625–631.
2. De Block, M., Botterman, J., Vandewiele, M., Dockx, J., Thoen, C., Gossele, V., Rao Movva, N., Thompson, C., Van Montagu, M., and Leemans, J. (1987) Engineering herbicide resistance in plants by expression of a detoxifying enzyme. *The EMBO Journal* 6, 2513–2518.
3. Thompson, C. J., Movva, N. R., Tizard, R., Crameri, R., Davies, J. E., Lauwereys, M., and Botterman, J. (1987) Characterization of the herbicide-resistance gene *bar* from *Streptomyces hygroscopicus. The EMBO Journal* 6, 2519–2523.
4. Gordon-Kamm, W., Spencer, T. M., Mangano, M. L., Adams, T. R., Daines, R. J., Start, W. G., O'Brien, J. V., Chambers, S. A., Adams, W. R. Jr., , Willetts, N. G., Rice, T. B., Mackey, C. J., Krueger, R. W., Kausch, A. P., and Lemaux, P. G. (1990) Transformation of maize cells and regeneration of fertile transgenic plants. *Plant Cell* 2, 603–618.
5. Brettschneider, R., Becker, D., and Lorz, H. (1997) Efficient transformation of scutellar tissue of immature maize embryos. *Theoretical and Applied Genetics* 94, 737–748.
6. Frame, B. R., Zhang, H., Cocciolone, S. M., Sidorenko, L. V., Dietrich, C. R., Pegg, S. E., Zhen, S., Schnable, P. S., and Wang, K. (2000) Production of transgenic maize from bombarded type II callus: Effect of gold particle size and callus morphology on transformation efficiency. *In Vitro Cellular Development Biology Plant* 36, 21–29.
7. Strauch, E., Wohlleben, W., and Puhler, A. (1988) Cloning of a phosphinothricin *N*-acetyltransferase gene from *Streptomyces viridochromogenes* Tu494 and its expression in *Streptomyces lividans* and *Escherichia coli. Gene* 63, 65–74.
8. Wehrmann, A., Vliet, A. V., Opsomer, C., Botterman, J., and Schulz, A. (1996) The similarities of *bar* and *pat* gene products make them equally applicable for plant engineers. *Nature Biotechnology* 14, 1274–1278.
9. Wohlleben, W., Arnold, W., Broer, I., Hillemann, D., Strauch, E., and Puhler, A. (1988) Nucleotide sequence of the phosphinothricin *N*-acetyltransferase gene from *Streptomyces viridochromogenes* Tu494 and its expression in *Nicotiana tabacum. Gene* 70, 25–37.
10. Murakami, T., Anzai, H., Imai, S., Satoh, A., Nagaoka, K., and Thompson, C. J. (1986) The bialaphos biosynthetic genes of *Streptomyces hygroscopicus*: Molecular cloning and characterization of the gene cluster. *Molecular and General Genetics* 205, 42–53.
11. Droge, W., Broer, I., and Puhler, A. (1992) Transgenic plants containing the phosphinothricin-*N*-acetyltransferase gene metabolize the herbicide I-phosphinothricin (glufosinate) differently from untransformed plants. *Planta* 187, 142–151.
12. Schwartz, D., Berger, S., Heinzelmann, E., Muschko, K., Welzel, K., and Wohlleben, W. (2004) Biosynthetic gene cluster of the herbicide phophinothricin tripeptide from *Streptomyces viridochomogenes* Tu494. *Applied and Environmental Microbiology* 70, 7093–7102.
13. Wohlleben, W., Alijah, R., Dorendorf, J., Hillemann, D., Nussbaumer, B., and Pelzer, S. (1992) Identification and characterization of phosphinothricin-tripeptide biosynthetic genes in *Streptomyces viridochromogenes. Gene* 115, 127–132.

14. Hoerlein, G. (1994) Glufosinate (phosphinothricin), a natural amino acid with unexpected herbicidal properties. *Reviews of Envionmental Contamination and Toxicology* 138, 73–145.
15. Frame, B. R., Shou, H., Chikwamba, R. K., Zhang, Z., Xiang, C., Fonger, T. M., Pegg, S. E. K., Li, B., Nettleton, D. S., Pei, D., and Wang, K. (2002) *Agrobacterium tumefaciens*-mediated transformation of maize embryos using a standard binary vector system. *Plant Physiology* 129, 13–22.
16. Zeng, P., Vadnais, D. A., Zhang, Z., and Polacco, J. C. (2004) Refined glufosinate selection in *Agrobacterium*-mediated transformation of soybean [*Glycine max* (L.) Merrill]. *Plant Cell Reports* 22, 478–482.
17. Duke, S. O. (2005) Taking stock of herbicide-resistant crops ten years after introduction. *Pest Management Science* 61, 211–218.
18. Dennehey, B. K., Retersen, W. L., Ford-Santino, C., Pajeau, M., and Armstrong, C. L. (1994) Comparison of selective agents for use with the selectable marker gene *bar* in maize transformation. *Plant Cell, Tissue and Organ Culture* 36, 1–7.
19. Zhao, Z., Lowe, K., and Marsh, W. (1993) *Bar* gene as a selection marker for maize transformation. *Maize Genetics Cooperation Newsletter* 67, 54.
20. Ahrens, W. (Ed.) (1994) *Herbicide Handbook.* Weed Science Society of America, Champaign, IL.

Chapter 13

Characterizing Transgene Inheritance

Earl H. Bicar

Summary

The analysis of transgene inheritance is an important step in the molecular and genetic characterization of transgenes. In this manuscript, two approaches to characterize the inheritance of transgenes are described. The first approach is based on the expression of the transgene phenotype and the second is based on the analysis of transgene DNA. Instructions on how to make crosses and develop breeding populations are outlined and the importance of these breeding populations in the analysis of transgene inheritance is explained. The number of individuals needed to determine segregation ratios and the statistic used to test these ratios are described. Examples of inheritance patterns that deviate from known expectations are provided and the possible causes of these deviations are discussed.

Keywords: Transgene, Porcine α-lactalbumin, Single locus, Chi-square test, Gene silencing

1. Introduction

Recent advances in genetic engineering and the improvements in transformation technology have provided new opportunities for plant improvement through the use of transgenes. Transgenes that express plant proteins have been successfully used to alter the amino acid balance of several crop species *(1–4)*. Transgenic varieties that confer resistance to insect pests and to nonselective herbicides for superior weed control have been produced *(5)*.

To be useful in breeding, it is important that transgenes are inherited in a stable and predictable manner. Transgenes that are inherited according to known expectations allow breeding procedures such as the integration of the transgene into elite germplasm by backcrossing and the evaluation of the trait across genetic backgrounds and environments in replicated trials.

M. Paul Scott (ed.), *Methods in Molecular Biology: Transgenic Maize, vol. 526*

DOI: 10.1007/978-1-59745-494-0_13

In this manuscript, the inheritance pattern of the transgene porcine α-lactalbumin is used as an example. α-Lactalbumin is a calcium-binding milk protein. α-Lactalbumin transgenes have been produced in maize using a synthetic coding sequence coupled to the maize ubi-1 promoter *(3)*. Changes in amino acid composition including high levels of lysine in the maize grain as a result of α-lactalbumin expression have been reported *(4)*.

The objectives of this paper are to *(1)* outline the methods and procedures used to determine the inheritance patterns of transgenes, *(2)* provide results on inheritance patterns that deviate from known expectations, and *(3)* describe and explain the possible causes of these deviations.

2. Materials

2.1. Pollination and Harvesting of the Segregating Populations

1. Lawson paper bags and clips to cover corn tassels a day before pollination and glacine bags to cover the silks and prevent contamination during silk emergence.
2. Mesh bags to contain harvested individual ears. Ear envelopes to store shelled kernels from individual corn ears.

2.2. Preparation of Samples for α-Lactalbumin Assay

1. Hand-drill to extract endosperm powder from individual kernels (10 mg).
2. Sample extraction buffer (10×): 0.5 M Tris–HCl pH 6.8; 10% SDS; 10% glycerol; 5% BME. Store at room temperature.

2.3. Western Blot for α-Lactalbumin

1. Set up buffer and transfer buffers.
2. Nylon-backed nitrocellulose membrane (0.45 μm) from Millipore and chromatography paper from Whatman.
3. PBS–Tween buffer (8.0-g NaCl; 0.2-g KCL; 1.15-g Na_2HPO_4; 0.24-g KH_2PO_4; 1-ml Tween 20).
4. Blocking buffer: 1% ovalbumin in PBS–Tween.

2.4. PCR for α-Lactalbumin DNA Sequences

1. Puregene DNA isolation kit (Gentra Systems).
2. DNA suspension buffer (50 mM Tris-HCL, 10 mM EDTA, pH 8.0).
3. PCR reaction volume (20 μl total): 100 μM of each dNTP, 2 μl of 10× PCR buffer (Gibco BRL), 1 unit Platinum Taq DNA Polymerase (Gibco BRL), 2.85 mM of $MgCl_2$, and 0.2 μM of each primer. Add 1 μl of isolated genomic DNA (~50 ng).
4. RapidCycler (Idaho Technologies).
5. 1.5% agarose gels stained with ethidium bromide.

3. Methods

Transgene inheritance usually is scored by observing the segregation of the transgene phenotype. In most cases, transgene DNA is inherited as a single Mendelian factor *(6)*. This implies the presence of only one integration locus or the presence of closely linked loci segregating together *(6)*. Segregation distortion has been observed resulting from abnormal expression of the transgene. In many instances this abnormal expression is detected in the first generation of transgenic plants while in some experiments the abnormal expression is not observed until large-scale field trials *(7)*. Therefore, long-term studies through several generations are required to characterize the inheritance of transgenes. Breeding populations are needed to study and determine whether the transgene is expressed, transmitted to the next generation, and inherited according to known expectations.

3.1. Development of Breeding Populations

3.1.1. F1 Populations

1. The T0 plants described in these instructions are the first plants obtained after transformation and subsequent regeneration in selective media. The T0 plants can be grown in jiffy pots in the glasshouse and then transplanted and grown in the field.
2. The F1 population is produced by matings between a positive T0 plant and a nontransgenic inbred line using the T0 plants as female (*see* **Note 1**) and the inbred line as male.
3. The BC1F1 population is produced by crossing positive F1 plants to a nontransgenic inbred line using the inbred as female and the positive F1 plants as male (*see* **Note 2**).

3.1.2. F2 and F3 Populations

1. F2 populations are developed by self-pollinating positive F1 plants, and F3 populations are developed by self-pollinating positive F2 plants. Self-pollination is performed by taking pollen from the tassel and then pouring the pollen into the silks of the same plant.
2. The silks of plants used as females are covered with glacine bags 2–3 days before the silks emerge in order to avoid contamination from unwanted pollen. Also, the tassels of plants used as pollen source are covered with paper bag to avoid contamination and to obtain fresh pollen on the day of pollination.
3. Depending on the maturity of the inbred line, ears can be harvested from 95 to 120 days after planting. In each breeding population, the harvested ears should be maintained as a distinct family by shelling them separately.

3.2. Characterizing Inheritance Patterns: Evaluation of the Transgene Phenotype

The α-lactalbumin phenotype is assessed by extracting the protein from the endosperm and performing a Western blot analysis to determine the presence or absence of the α-lactalbumin protein in the kernels (*see* **Note 3**).

3.2.1. Detecting α-Lactalbumin by Western Blot

1. A small portion of the endosperm (~10 mg) from the kernels is removed by using a hand-held drill. Proteins are extracted with 100 μl of SDS–PAGE sample buffer (0.5 M Tris–HCl pH 6.8; 10% SDS; 10% glycerol; 5% BME) per 10 mg of endosperm powder.
2. The samples are then placed in a vortex shaker for 30 min. The insoluble material is removed by centrifugation at 13,000 rpm for 5 min in a microcentrifuge. The supernatant is pipetted into 0.5 ml appropriately labeled Eppendorf tubes.
3. The supernatant is boiled for 5 min before loading on a 15% SDS-PAGE gel. The PAGE gels are blotted onto a nylon-backed nitrocellulose membrane (0.45 μm) using a minitrans-blot apparatus (Bio-Rad).
4. The membrane is treated with 1% ovalbumin in PBS–Tween buffer (8.0-g NaCl; 0.2-g KCL; 1.15-g Na_2HPO_4; 0.24-g KH_2PO_4; 1-ml Tween 20) for 1 h and allowed to react for 8–10 h with a polyclonal antibody against human α-lactalbumin raised in rabbits and then visualized according to the manufacturer's protocol for colorimetric visualization of an alkaline phosphatase-conjugated anti-rabbit IgG (Bio-Rad).
5. The number of kernels that are positive and negative for α-lactalbumin is recorded.

3.2.2. Number of Kernels to Test

In transgene segregation analysis, the number of kernels needed in order to evaluate the transgene phenotype is important. The number of kernel samples (n) to evaluate depends on the breeding generation and the expected frequency of positive kernels in that generation (q), the number of kernels desired that are positive for the transgene (r), and the intended probability of success (p). For example, it is expected that an F1 ear produced from a cross between a positive T0 plant and a null inbred will contain 50% positive kernels ($q = ½$). If 5 positive kernels are desired ($r = 5$) with 95% probability of success ($p = 0.95$), a minimum of 16 F1 kernels need to be evaluated. **Table 1** summarizes the number of individuals (i.e., plants or kernels) to evaluate in order to obtain the desired number containing the transgene of interest.

3.2.3. Segregation Analysis in the F1 and BC1F1: Test for 1:1 Segregation

1. F1 and BC1F1 kernels are evaluated for α-lactalbumin by Western blot as described in **Subheading 3.2.1**. The number of kernels that are positive and negative for α-lactalbumin is recorded and the ratio of positive to negative kernels is determined.

Table 1
Number of plants necessary to recover a required number of plants with trait *(8)*

		r = number of plants to be recovered								
	q	1	2	3	4	5	6	8	10	15
p = 0.95	3/4	3	5	6	8	9	11	14	17	25
	1/2	5	8	11	13	16	18	23	28	40
	1/3	8	13	17	21	25	29	37	44	62
	1/4	11	18	23	29	34	40	50	60	84
	1/8	23	37	49	60	71	82	103	123	172
	1/16	47	75	99	122	144	166	208	248	347
	1/32	95	150	200	246	291	334	418	500	697
	1/64	191	302	401	494	584	671	839	1,002	1,397
		r = number of plants to be recovered								
	q	1	2	3	4	5	6	8	10	15
p = 0.99	3/4	4	6	8	9	11	13	16	19	27
	1/2	7	11	14	17	19	22	27	33	45
	1/3	12	17	22	27	31	35	44	52	71
	1/4	17	24	31	37	43	49	60	70	96
	1/8	35	51	64	77	89	101	124	146	198
	1/16	72	104	132	158	182	206	252	296	402
	1/32	146	210	266	318	368	416	508	597	809
	1/64	293	423	535	640	739	835	1,020	1,198	1,623

p = probability of recovering. *r* plants with trait. *q* = probability of occurrence of trait

2. Chi-square analysis is then performed to test whether the observed ratio of positive to negative kernels fits with the expected phenotypic ratio for a single dominant locus segregating in the F1 (*see* **Note 4**). As described in **Subheading 3.1.1**, the F1 population was produced by matings between a T0 plant positive for the α-lactalbumin transgene and a nontransgenic inbred. The T0 plant is hemizygous for the transgene and so a 1:1 positive to negative phenotypic ratio is expected among kernels in the F1 (*see* **Note 5**). An example of the segregation ratio of F1 kernels expressing α-lactalbumin is shown in **Table 2**.
3. The BC1F1 population was produced by matings between positive F1 plants and nontransgenic inbred using the F1 plants as males as described in **Subheading 3.1.1**. The positive F1 plants

are heterozygous at the transgene locus and carry one copy of the transgene. Therefore a 1:1 positive and negative phenotypic ratio is expected among kernels in a backcross population (*see* **Note 6**). An example of the segregation ratio of BC1F1 kernels expressing α-lactalbumin is shown in **Table 2**.

4. Outlined here is an illustration on how to perform a chi-square analysis or test for goodness of fit. We will use the BC1F1 data from event P45-22 in **Table 2** summarized here:

	Positive	Negative
Observed	4	6
Expected	5	5

Table 2
Segregation of a modified α-lactalbumin transgene among independent events in maize based on the presence and absence of the α-lactalbumin protein as detected by Western blot analysis *(9)*

Event number	Generations			
	F1 (+/−)	BC1F1 (+/−)	F2 (+/−)	F3 (+/−)
P45-3	2/3	3/7 ns	34/10 ns	1/16**; 8/9**; 2/15**; 3/14**
P45-22	4/1	4/6 ns	43/20 ns	8/9**; 11/6 ns; 8/9**; 1/16**
P45-16	4/1	3/7 ns	53/11 ns	17/0 H; 17/0 H; 17/0H
P45-11	3/2	1/9*	33/28**	1/16**; 8/9**; 2/15**; 3/14**
P45-4	2/3	0/10**	0/44**	3/14**; 0/17**; 2/15**; 5/12**; 0/17**
P45-19	3/2		0/44*	1/16**; 5/12**
P57-35	3/2	0/10**	0/40**	0/17**; 0/17**; 0/17**; 1/16**; 0/17**
P57-24	3/2	0/10**	0/61**	1/17**; 0/17**; 2/15**; 0/17**; 0/17**
P57-27	5/0	0/10**	0/61**	
P57-25	3/2	0/10**	0/40**	

ns not significantly different from 3:1 phenotypic ratio for a single dominant locus model in the F2 and F3; 1:1 ratio for BC1F1

**Significantly different; χ^2 (0.01, 1df) = 6.64

*Significantly different; χ^2 (0.05, 1df) = 3.84 *H* homozygous for α-lactalbumin protein expression. Numbers in each generation column are the proportion of kernels expressing (+)/not expressing (−) the α-lactalbumin protein in the endosperm. F1 kernels were produced by crossing a T0 plant and B73. Plants resulting from these kernels were crossed to B73 or self-pollinated to get BC1F1 or F2 kernels, respectively. F2 plants were self-pollinated to get F3 kernels. In P45, the synthetic α-lactalbumin coding sequence is transcriptionally regulated by the maizeUbi-1 promoter and the nos 3′ untranslated region. The coding region encodes a mature porcine α-lactalbumin and is translationally fused to the 27-kDa gamma zein signal sequence at the N-terminus and an ER retention signal at the C-terminus of the protein. In P57, the synthetic α-lactalbumin coding sequence is transcriptionally regulated by the maize 27-kDa gamma zein promoter and the nos 3′ untranslated region. The coding sequence does not encode an ER retention signal at the C-terminus of the protein

1. *Establish the hypothesis.* We want to test using a chi-square statistic the null hypothesis (H_0) that our observed segregation ratio of 4 positive:6 negative plants in the BC1F1 population is not significantly different from the phenotypic 5:5 ratio (or 1:1) expected for a single dominant locus segregating in the F1 generation.
2. *Calculate for the χ^2 statistic.* We calculate the χ^2 using the formula $\chi^2 = \Sigma[(O - E)^2/E]$ (*see* **Note 4**) and substitute the values for O and E.
   (a) For Positives: $\chi^2 = (O - E)^2/E = (4 - 5)^2/5 = 0.20$
   (b) For Negatives: $\chi^2 = (O - E)^2/E = (6 - 5)^2/5 = 0.20$ 0000The sum of these categories is $\chi^2 = 0.40$.
3. *Assess for the significance levels.* The significance of the χ^2 value is determined by comparing the χ^2 value with the critical chi-square value. The critical chi-square value is obtained by calculating the degrees of freedom (df) and by using the chi-square distribution table. The degrees of freedom is obtained by the number of categories – 1. The critical chi-square value is obtained by looking at the value in the chi-square distribution table that corresponds to the df value. The chi-square distribution table is found in most statistics textbooks *(10, 11)*. For our example, the df = 2 – 1 = 1 and $\alpha = 0.01$. The critical value of chi-square is as follows:

$$\chi^2_{(0.01,1df)} = 6.64.$$

We will reject the null hypothesis (H_0) if the calculated value of χ^2 statistic is bigger than χ^2 critical value. Since the calculated value of the test statistic, $\chi^2 = 0.40$, is less than $\chi^2_{(0.01,\ 1df)} = 6.64$, the null hypothesis ($H_0$) can not be rejected. There is insufficient information to indicate a lack of fit of the observed segregation ratio to the expected ratio. We conclude that the observed 4:6 ratio is not significantly different from the expected 1:1 segregation ratio for a single dominant locus segregating in the BC1F1 (*see* **Note 7**).

3.2.4. Segregation Analysis in the F2: Test for 3:1 Segregation

1. F2 kernels are evaluated for α-lactalbumin as described in **Subheading 3.2.1**. The number of kernels that are positive and negative for α-lactalbumin in a Western blot is again recorded.
2. Chi-square analysis is then performed as described in **Subheading 3.2.3** to test whether the observed ratio of positive:negative kernels fits with the expected phenotypic ratio for a single dominant locus segregating in the F2.
3. As described in **Subheading 3.1.2**, the F2 population was produced by self-pollination of positive F1 plants that are

heterozygous at the transgene locus. Therefore a 3:1 positive to negative phenotypic ratio for α-lactalbumin is expected in F2 kernels (*see* **Notes 8** and **9**). An example of the segregation ratio of F2 kernels expressing α-lactalbumin is shown in **Table 2**.

3.2.5. Segregation Analysis in the F3: Test for Zygosity

1. F3 kernels are evaluated for α-lactalbumin as described in **Subheading 3.2.1** and the number of kernels that are positive and negative for α-lactalbumin in a Western blot is again recorded.
2. Chi-square analysis is again performed as described in **Subheading 3.2.3** to test whether the observed ratio of positive:negative kernels expressing α-lactalbumin fits the expected phenotypic ratio for a single dominant locus segregating in the F3.
3. The F3 population is produced by self-pollination of an F2 plant positive for α-lactalbumin transgene as described in **Subheading 3.1.2**. A 3:1 phenotypic ratio for α-lactalbumin is expected in F3 kernels derived from selfing a heterozygous F2 plant while 100% expression of α-lactalbumin is expected in F3 kernels derived from selfing a homozygous F2 plant (*see* **Note 10**). An example of the segregation ratios of F3 populations expressing α-lactalbumin is shown in **Table 2**.

3.3. Characterizing Inheritance Patterns: DNA-Based Analysis

1. The inheritance patterns of transgenes can also be determined by DNA-based approaches such as the use of conventional or real-time PCR techniques.
2. In a conventional PCR, for example, transgene or transgene DNA sequences can be detected in transgenic plants or kernels using primers that correspond to the transgene coding sequence. The presence or absence of a PCR band would indicate that the kernel or plant is positive or negative for the transgene.
3. Segregation analysis is then performed in the F1, F2, and F3 generations as described in **Subheading 3.2.3** to determine transgene segregation patterns (*see* **Note 11**).

3.4. What to Do When Abnormal Segregation Ratios Are Detected?

1. When abnormal transgene segregation ratios are detected, it is important to understand and determine the factors that contributed to the distortion.
2. If abnormal transmission of the transgene is suspected, reciprocal crosses can be performed to determine whether the abnormal transmission was from the male or female gamete as described in **Note 8**.
3. In cases when the segregation distortion is due to an abundance of negative kernels (null phenotype), DNA-based

analysis such as a PCR should be performed to determine the presence or absence of the transgene on the negative kernels or plants. PCR analysis conducted on the negative kernels such as in events P45-11, P45-4, P57-35, and P57-24 in **Table 2** showed that α-lactalbumin DNA sequences were present in these null kernels.

4. Notes

1. In many cases, due to tissue culture conditions, T0 plants produce poor ears and limited pollen or do not produce pollen at all. If the T0 plant produces pollen, it is best to use the positive T0 plants as the male parent in producing the F1 population in order to maximize seed set.
2. Corn is monoecious with male and female flowers on the same plant. In these instructions, male plants refer to those plants used as the source of pollen. Female plants refer to those plants whose silks are fertilized by pollen and where the ears are harvested. In making the BC1F1 population, the positive F1 plants are used as males in the cross in order to introduce the transgene into a normal cytoplasm.
3. Some transgene phenotypes are easier to assess. For example, in YieldGard corn that confers resistance to European corn borer, the Cry1Ab protein is detected readily by using dipsticks (EnviroLogix, Inc). In Roundup Ready corn that confers tolerance to glyphosate, the phenotype is assessed by spraying corn plants with Roundup herbicide (*see* the chapter "Herbicide Resistance Screening Assay"). An alternative to western blotting for immunological detection of transgenes is presented in the chapter "Tissue-Print Immunodetection of Transgene Products in Endosperm for High-Throughput Screening of Seeds."
4. The chi-square statistic is used to determine if a distribution of observed frequencies differs from the theoretical expected frequencies. The value of the chi-square statistic is obtained by the formula: $\chi^2 = \Sigma[(O - E)^2/E]$ where χ^2 is the chi-square statistic, O is the observed frequency, and E is the expected frequency *(10, 11)*. The chi-square test for goodness of fit compares the expected and observed values in order to determine how well a statistical model fits a set of observations. In our analysis of transgene segregation, we want to test using a chi-square statistic the null hypothesis that our observed segregation ratio of positive:negative plants in the F1 populations is not significantly different from the expected

phenotypic 1:1 ratio for a single dominant locus segregating in the F1 generation.

5. Because the transgenic locus is, in most cases, hemizygous and most transgenes provide plants with a gain of function, they behave as dominant genes, and a 1:1 segregation ratio for transgenic to wild-type progeny is normally expected in diploids or diploidized polyploids when a hemizygous plant is mated with a nontransgenic plant. Since the T0 plant was used as the female in the cross, a segregation ratio that fits the expected 1:1 ratio in the F1 also indicates the normal transmission of the transgene through the female gamete.
6. The F1 plant was used as the male in the cross with a nontransgenic inbred to form the BC1 population, and so a segregation ratio that fits the expected 1:1 ratio in the BC1F1 indicates normal transmission of the transgene through the pollen. This form of testcross can be repeated for several generations by continuously crossing positive F1 plants to unrelated nontransgenic inbreds. A consistent 1:1 segregation ratio will not only indicate normal transmission but also stability of the transgene across generations. Also, since the segregation analysis was performed by the analysis of the transgene phenotype by Western blot, the results demonstrated that not only was the α-La transgene transmitted to the BC1F1 generation but that the transgene functioned to produce immunologically detectable α-lactalbumin in the F1 kernel endosperm.
7. Because it is widely used, the chi-square test is also one of the most abused statistical procedures. The user should always be certain that the experiment satisfies the properties of a multinomial experiment before proceeding with the test *(11)*. In genetic experiments, the chi-square test must be used only on numerical data itself and not on percentages or ratios derived from the data *(12)*. Also, the chi-square test should be avoided when the estimated expected count (or expected frequency within any phenotypic class) is small, for it is in this instance that the chi-square probability distribution gives poor approximation to the sampling distribution of the χ^2 statistic. As a rule of thumb, an estimated expected count of at least 5 will mean that the chi-square distribution can be used to determine an approximate critical value of χ^2 *(11)*.
8. A 3:1 phenotypic ratio is normally expected in diploids or diploidized polyploids when a heterozygous F1 plant is selfed. This segregation ratio is obtained when the transgene is inherited as single dominant locus. Such ratio also indicates normal transgene transmission in both male and female gametes. Distortion of such segregation pattern has

been reported in F2 populations of transgenic maize as a result of reduced pollen transmission due to reduced pollen viability *(13, 14)*. As a result of reduced transmission of the transgene through the pollen, a 1:1 segregation ratio was observed in the F2 rather than 3:1. Abnormal segregation ratios due to reduced transmission of the transgene through the male gamete can be further verified by making primary and reciprocal crosses with nontransgenic inbreds and comparing the segregation ratios observed in both crosses. In a primary cross, the transgenic F1 is used as a female [e.g., positive F1 (female) × nontransgenic inbred (male)] while in a reciprocal cross the transgenic F1 is used as a pollen source [e.g., nontransgenic inbred (female) × positive F1 (male)]. Normal transmission of the transgene through the female gamete in the primary cross or through the pollen in the reciprocal cross should result in a 1:1 phenotypic ratio in both crosses.

9. Abnormal segregation ratios have also been reported as a result of abundance of kernels showing the negative or null phenotype *(9)*. For some events, the abundance of negative kernels was observed as early as in the F1 and was maintained in the F2 and F3 while for other events, the increase in null phenotypes manifested in the F3 (**Table 2**). PCR analysis of plants derived from the negative kernels revealed the presence of a full length α-lactalbumin transgene DNA sequences indicating that the transgene was inactivated in these negative kernels. This distortion of the segregation pattern as a result of transgene inactivation can often be explained by gene silencing. Transgene silencing herein defined as the inactivation of the transgene despite the presence of unchanged, but possibly methylated transgene sequence in the plant genome *(15)* has been extensively documented in transgenic plants produced by *Agrobacterium*-mediated transformation and also those transformed by microprojectile bombardment *(16, 17)*. Presence of sequences homologous to either transgene sequences or endogenous gene sequences has been reported to be associated with transgene inactivation. Homologous sequences can lead to methylation of the promoter region blocking transcription *(18)*. Alternatively, one model also postulates that such methylated homologous sequences may form or recruit repressive chromatin complexes. The repressive chromatin then spreads to adjacent homologous sequences or interacts in trans with other homologous sequences altering the chromatin structure *(16, 17)*.

10. Classification of F2 kernels into positives and negatives was based on the expression of the transgene phenotype by

Western blotting. Positive F2 kernels can either be heterozygous or homozygous for the transgene. In this phenotypic assay, the homozygous and heterozygous are lumped into one positive class. Since the positive kernels are lumped into one class, the heterozygous and homozygous F2 kernels can only be identified by progeny testing. When F2 plants are grown, F3 kernels derived from selfing a heterozygous F2 plant will segregate 3:1. F3 kernels derived from selfing a homozygous F2 plant will have all kernels expressing α-lactalbumin in the endosperm.

11. In a conventional PCR-based segregation analysis, a 3:1 positive to negative ratio is expected for a transgene segregating as a single dominant locus in the F2. Similar to the phenotypic assay described in **Subheading 3.2.4**, the positive homozygotes and heterozygotes are also lumped into one positive class. Since the homozygotes and heterozygotes are lumped into one positive class, a progeny test is necessary to identify the positive individuals homozygous for the transgene. When the objective is to rapidly identify homozygous plants, progeny testing is a time-consuming and laborious practice. Real-time PCR is another DNA-based method that can be used to assess transgene segregation patterns. Real-time PCR does not only detect presence of the transgene but also quantifies the amount of amplified DNA. It permits accurate comparison of DNA amounts and allows clear discrimination between twofold DNA levels. It can therefore accurately distinguish homozygous and heterozygous transgenic plants *(19)*. With real-time PCR segregation analysis, homozygous and heterozygous classes can be clearly distinguished. For a transgene segregating as a single dominant locus in the F2, a 1:2:1 ratio for individuals homozygous for the transgene, individuals heterozygous for the transgene, and individuals homozygous for no transgene (null) is expected. Real-time PCR has been used accurately to analyze zygosity of transgenic plants *(19, 20)*.

References

1. Lai, J. S. and Messing, J. (2002). Increasing maize seed methionine by mRNA stability. *Plant J.* 30, 395–402.
2. Chakraborty, S., Chakraborty, N. and Datta, A. (2000). Increased nutritive value of transgenic potato by expressing a nonallergenic seed albumin gene from Amaranthus hypchondriacus. *Proc. Natl. Acad. Sci. USA* 97, 3724–3729.
3. Yang, S. H., Moran, D. L., Jia, H. W., Bicar, E. H., Lee, M. and Scott, M. P. (2002). Expression of a synthetic porcine α-lactalbumin gene in the kernels of transgenic maize. *Transgenic Res.* 11, 11–20.
4. Bicar E.H., Woodman-Clikeman W., Sangtong V., Peterson J.M., Yang S.S., Lee M., and Scott M.P. (2008). Transgenic maize endosperm containing a milk protein has improved amino acid balance. *Transgenic Res.* 17, 59–71.
5. Crosbie T. M et al. (2006). Plant Breeding: Past, Present and Future. In: *Plant Breeding: The Arnel Hallauer International Symposium.*

(K.R. Lamkeyand M. Lee, eds). Blackwell, Ames, Iowa, pp. 3–50.

6. Pawlowski W. P. and Somers D. A. (1996). Transgene inheritance in plants genetically engineered by microprojectile bombardment. *Bio/Technology*. 6, 17–30.
7. Armstrong C.L., Parker G., Pershing J. et al. (1995). Field evaluation of European corn borer control in progeny of 173 transgenic corn events expressing an insecticidal protein from Bacillus thuringiensis. *Crop Sci.* 35, 550–557.
8. Sedcole, J.R. (1977). Number of plants necessary to recover a trait. *Crop Sci.* 17, 667–668.
9. Bicar E.H. (2003). Inheritance and expression of porcine α lactalbumin transgenes. In: *Molecular and genetic characterization of porcine α lactalbumin transgenes*. Ph.D. Dissertation. Iowa State University.
10. Steel R.D. and Torrie J.H.(1960). *Principles and Procedures of Statistics.* McGrawhill, New York.
11. Sincich T. (1993). Categorical data analysis. In: Statistics by Example. Dellan-MacMillan, New York, pp. 809–817.
12. Stansfield W.D. (1983). *Statistical distributions.* In: *Theory and Problems in Genetics.* McGraw-Hill, New York, pp. 140–155
13. Sangtong V., Moran D., Chikwamba R., Wang K., Woodman W., Lee M., and Scott M. (2002). Expression and inheritanace of the wheat Glu-1DX5 gene in transgenic maize. *Theor. Appl. Genet.* 105, 937–945.
14. Scott M.P., Peterson J.M., Sangtong V., Moran D., Smith L. (2007). A wheat genomic DNA fragment reduces pollen transmission in maize transgenes by reducing pollen viability. *Transgenic Res* 16, 629–643.
15. Pawlowski W.P., Tolbert K.A., Rines H.W., and Somers D.A. (1998). Irregular patterns of transgene silencing in allohexaploid oat. *Plant Mol. Biol.* 38, 597–607.
16. Chandler V.L. and Vaucheret H. (2001). Gene inactivation and gene silencing. *Plant Physiol.* 125, 145–148.
17. Fagard M. H. Vaucheret. (2000). (Trans) Gene silencing in plants: How many mechanisms? *Annu. Rev. Plant. Physiol. Plant. Mol. Biol.* 51, 167–194.
18. Miitelsen S. Afsar K. and Paszkowski J. (1998). Release of epigenetic gene silencing by transacting mutations in Arabidopsis. *Proc. Natl. Acad. Sci.* 95, 632–37.
19. German MA., Kandel-Kfir, M., Swarzberg, D.,Matsevitz, T., Garnot, D. (2003). A rapid method for the analysis of zygosity in transgenic plants. *Plant Sci.* 164, 183–187.
20. Bubner B. and I.T. Baldwin. (2004). Use of real-time PCR for determining copy number and zygosity in transgenic plants. *Plant Cell Rep.* 23, 263–271.

Chapter 14

Backcross Breeding

Karla E. Vogel

Summary

Backcross breeding enables breeders to transfer a desired trait such as a transgene from one variety (donor parent, DP) into the favored genetic background of another (recurrent parent, RP). If the trait of interest is produced by a dominant gene, this process involves four rounds of backcrossing within seven seasons. If the gene is recessive, this process requires more generations of selfing and thus nine or more seasons are needed. The rate at which the DP genes are removed and the RP genes are recovered in the genetic makeup of the plant can be calculated using the number of backcross generations utilized. This rate is dramatically increased with the recent advances in marker technology which allow breeders to control the gene of interest and control the genetic background.

Keywords: Donor parent, Recurrent parent, Molecular markers

1. Introduction

Backcross breeding is a widely used method of transferring a desired trait, either natural or engineered, from one plant to the elite background of another. This process is widely used when working with transgenes because transformable maize lines often have inferior agronomic properties. The parent contributing the transgene is known as the donor parent (DP), and the agronomically superior parent to which the successive backcrosses are made is known as the recurrent parent (RP). The effect of these successive backcrosses is to minimize the genomic contribution of the DP. Typically breeders try to recover 98% of the RP genome in the resulting line. For the backcross method to be successful the trait transferred must retain effectiveness throughout several generations and a sufficient number of backcrosses must be made to

M. Paul Scott (ed.), *Methods in Molecular Biology: Transgenic Maize, vol. 526*

DOI: 10.1007/978-1-59745-494-0_14

recover the desirable traits of the RP *(1)*. In the end, the resulting line should express the trait conferred by the transgene and the agronomic superiority of the RP.

In 1922, Harlan and Pope first proposed the use of backcrossing to develop crop plants. They noted that backcrosses had been used for many years in animal breeding to confer fixed traits and conveyed that its value had not yet been appreciated in agriculture *(1)*. After conducting one round of backcrossing they proved that a portion of the resulting progeny had the phenotypic integrity of one parent and the desired genetics of the other. This gave breeders an effective and inexpensive way to get the desired trait into the preferred genetic background *(2)*.

2. Dominant Allele Backcrossing

Most transgenes are inherited as dominant genetic loci. The process of backcrossing for dominant genes is shown in **Fig. 1.** The example shown is for a single, dominant transgene. After the transgenic plant has been obtained, it is planted near the elite inbred line that has been chosen for its desirable phenotype. During the first season, pollen is collected from the transgenic inbred, or DP, and deposited on the silk of the elite inbred, or RP. The progeny of this cross are called the F1 population and all seeds are harvested. These progeny are now heterozygous for the gene of interest, and the genetic makeup is 50% of the DP and 50% of the RP.

In season two, the F1s are planted, again near the RP. The F1 is then crossed back, hence the term backcrossing, to the RP. The progeny from this cross are harvested at the end of the season and are known as the BC1F1 population. This population normally segregates equally into homozygous for the recessive, RP allele and heterozygous for the dominant, DP allele. Selection is usually made for the heterozygous plants by in-field assay. All plants homozygous for the recessive (RP) allele are discarded. The selected plants heterozygous for the gene of interest are now 75% RP and 25% DP genotype.

The seed from the BC1F1 population is backcrossed to the RP again in season three. The resulting population, named BC2F1, once again segregates equally into homoqygous recessive and heteroygous dominant for the gene of interest. All plants undergo an assay to identify plants homozygous for the gene of interest. The genotype of BC2F1 is now 87.25% RP and 16.25% donor.

In seasons four and five this process is repeated, giving rise to the BC3F1 and BC4F1 populations, respectively. At this point,

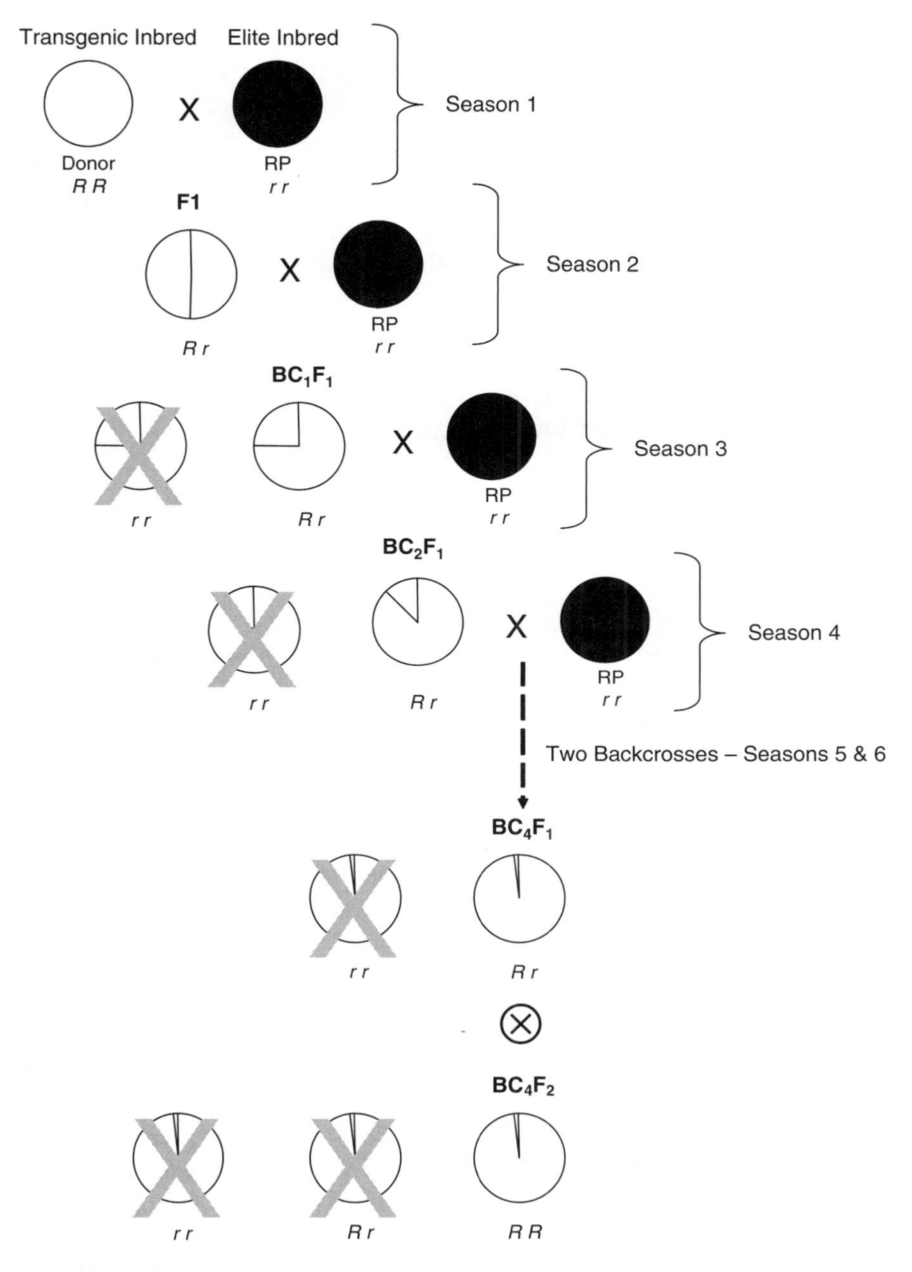

Fig. 1. Schematic representation of backcross breeding for a single, dominant transgene.

the 98% RP requirement has often been met. The heterozygous plants now contain most of the genome from the original elite line except for the gene of interest.

The objective of the breeding program is often a line that is homozygous for the dominant gene of interest. However, in a backcrossing scheme the transgene is propagated in a heterozygous state and it is necessary to pollinate the resulting line to achieve homozygosity. Thus, BC4F1 plants are selfed in season seven. The resulting progeny, BC4F2, will be ¼ homozygous for the gene of interest (RR), ½ heterozygous (Rr), and ¼ homozygous for the recessive allele (rr). This ratio can also be written as 1rr:2Rr:1RR. Since there is no way to distinguish Rr from RR when working with a dominant allele, assay or progeny testing must be used. Progeny testing is defined as breeding of the offspring to determine the genotypes of the parent. For example, selfing an RR BC4F2 will result in all progeny conveying the trait of interest. On the other hand, selfing of an Rr BC4F2 will give plants that segregate for the desired trait. Three quarters will demonstrate the dominant phenotype, but a fourth will not.

3. Recessive Allele Backcrossing

When dealing with a recessive gene, the process of backcrossing becomes more involved. The overall progression of backcrossing generations is the same, but seasons of backcrossing are separated by one or two seasons of self-pollination (*see* **Fig. 2**) to determine the plants carrying the recessive allele. A molecular genetic assay, rather than a phenotype assay, would shorten this time period.

As with dominant allele backcrossing, season one begins with the planting of the transgenic line, homozygous for the chosen recessive allele (rr), and the elite inbred, homozygous for the undesirable dominant allele (RR). The backcross is made to create the F1 population. All seeds are heterozygous for the recessive allele.

In season two, the F1 plants are crossed to the RP to obtain the BC1F1 generation. Fifty percent of these seeds are homozygous dominant (RR) and fifty percent are heterozygous (Rr).

In order to differentiate these two genotypic classes, the BC1F1 population is self-pollinated during the third season. The progeny are denoted BC1F2 and typically segregate 1rr:2Rr:1RR.

The BC1F2 seeds are planted in season four, and the homozygous recessive plants (rr) are identified through some type of phenotypic assay. These plants are then crossed back to the RP to obtain BC2F1 seeds. All seeds are heterozygous for the recessive allele (Rr).

In season five, the BC2F1 plants are crossed to the RP and yield BC3F1 seed. Fifty percent of these seeds are homozygous dominant (RR) and fifty percent are heterozygous (Rr).

The BC3F1 seed is planted in season six, and selfed. The progeny are identified as BC3F2.

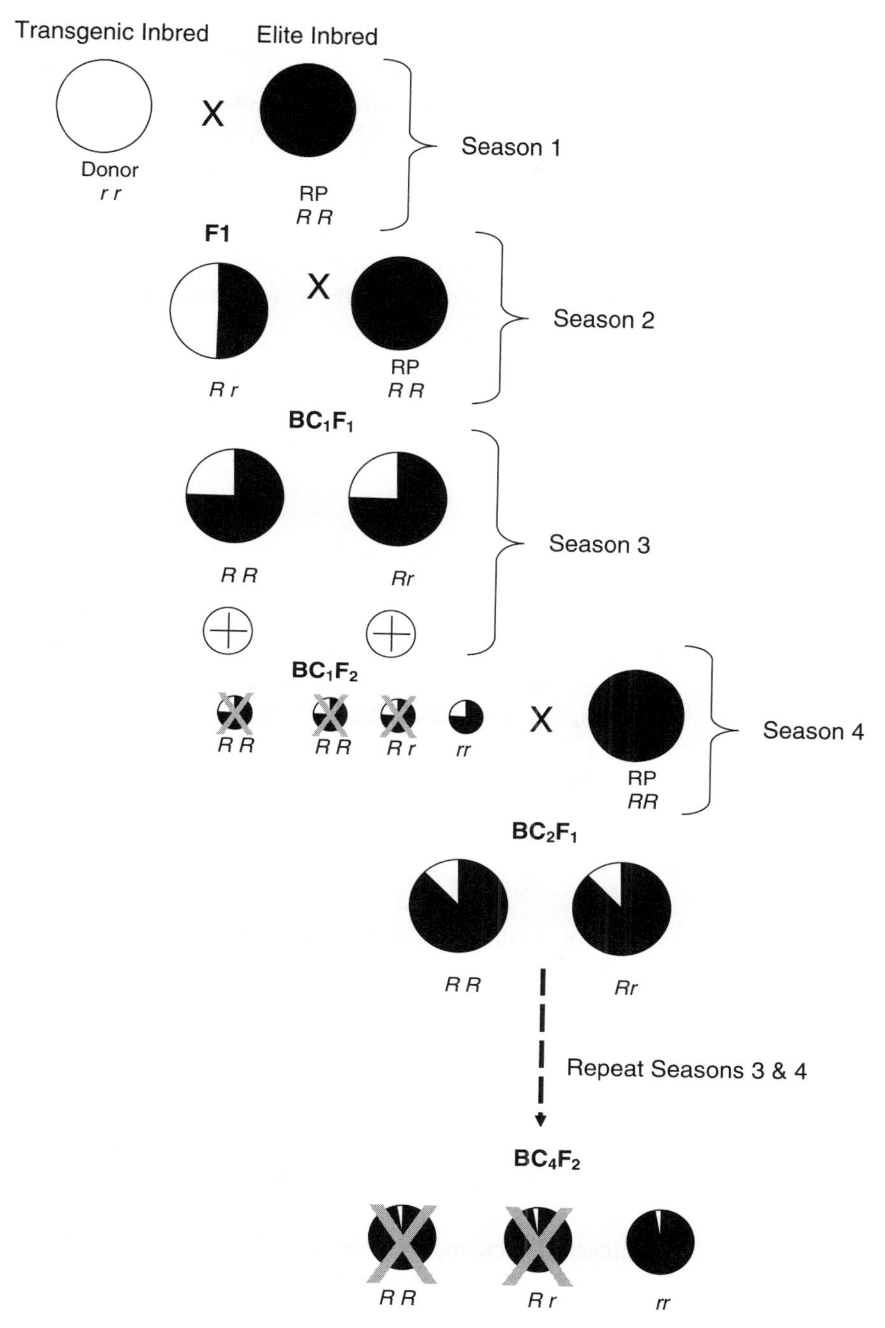

Fig. 2. Schematic representation of backcross breeding for a single, recessive transgene.

The BC3F2 seed is put through the same process as season four in season seven. In the end, the breeder then has a BC4F1 population heterozygous for the recessive allele (Rr).

In season eight, the BC4F1 population is planted and self-pollinated to obtain BC4F2 seeds. Once again, these seeds segregate 1RR:2Rr:1rr on average.

During season nine, the BC4F2 seed is planted and homozygous recessive plants (rr) are identified. Seeds from these plants are harvested and, in subsequent seasons, evaluated for desirable traits from the RP.

4. Recovery Rate of RP Genes

The overall goal of backcrossing is to recover all the genes of the RP except for the gene of interest. The extent of this recovery is dependent on the number of backcrosses done in the process and the number of loci that differ between the RP and the donor.

Assuming the absence of genetic linkage, the average recovery of RP genes increases each backcross by one-half the percentage of the DP present in the previous backcross. This is demonstrated in **Table 1.** The general equation is

$$\left(\frac{1}{2}\right)^{n+1} = \%\text{RP},$$

where n equals the number of backcrosses that have been completed *(2)*.

Table 1
Average recovery of RP genes per round of backcrossing assuming absence of gene linkage

Number of backcrosses	Recurrent parent (%)	Donor parent (%)
F1	50.00	50.00
1	75.00	25.00
2	87.50	12.50
3	93.75	6.25
4	96.88	3.13
5	98.44	1.56

Table 2
Average recovery of RP genome when recurrent parent and the donor parent have different alleles at multiple loci

Number loci	Backcross numbers					
	1 (%)	2 (%)	3 (%)	4 (%)	5 (%)	6 (%)
1	50.00	75.00	87.50	93.75	96.88	98.44
2	25.00	56.25	76.56	87.89	93.85	96.90
3	12.50	42.19	66.99	82.40	90.91	95.39
4	6.25	31.64	58.62	77.25	88.07	93.89
5	3.13	23.73	51.29	72.42	85.32	92.43
10	0.10	5.63	26.31	52.45	72.80	85.43

If the RP and the DP have different alleles at multiple loci the amount of backcrossing needed to obtain a high percentage of RP germplasm increases, as shown in **Table 2**. The general equation proposed by Allard in 1960 is

$$\left(\frac{(2^m - 1)}{2^m}\right)^n = \%\text{RP},$$

where m is the number of backcrosses and n is the number of loci that differ between the RP and DP *(3)*.

For example, if the DP and the RP have different alleles at ten loci only 85% of the BC6F1 plants will be homozygous for the ten alleles of the RP. In contrast, 98% of the BC6F1 plants will be homozygous for the RP allele if only one locus differs between the DP and the RP. It is for this reason that the DP should be as closely related to the RP as possible to reduce the number of backcross generations that need to be completed. With ten differing loci, nine backcrosses would need to be made to obtain the 98% homozygosity that is achieved in only six backcrosses when one allele is different.

5. Marker-Assisted Backcrossing

With the cost of producing transgenics and the importance of speed to market, the traditional approach of self-pollination in between backcrossing generations is too slow. Over the past 20 years many advances have been made in the field of molecular markers that have assisted breeding significantly. The ability to select desirable lines based on genotype rather than phenotype is an extremely important asset to plant breeders. This relatively new technology allows breeders to analyze plants at the seedling stage, and characterize multiple genetic loci at one time. Marker technology allows quicker turnaround of generations, cutting out the self-pollination generation.

In backcross breeding, molecular markers are utilized in two ways. First, the markers can be used to control the gene of interest. This is called foreground selection *(4)*. When using conventional backcrossing methods, integrating a recessive gene into an elite line requires additional seasons for self-pollination after each backcross. This prohibits the rapid advancement of lines that is normally needed in commercial breeding programs. When a phenotype of the desired gene cannot be easily assayed, but a marker is present within or near the target gene, foreground selection allows breeders to select heterozygous plants easily and rapidly often without the need for self-pollination *(5)*.

Second, molecular markers can be used to control the genetic background, a process known as background selection *(4)*. Backcross generations can be analyzed for the percentage of donor genome still present in the population using markers known in the elite line. Because the individual plants in a given generation of backcrossing contain different amounts of the RP genome, selections can be made for future generations of backcrossing using this genotypic data to increase the recovery rate of the RP genome. Visscher et al. *(6)* report a 98.1% RP recovery after only two backcrosses with the use of three genetic markers on a single chromosome. This is a 10.3% increase from conventional backcross breeding strategies which would take four backcross generations to achieve such high recovery of the RP genome.

To perform marker-assisted selection on the genetic background, markers of known map position are keys to comprehensive genomic coverage that drives effective selection for the RP. Genomic coverage is based on those markers that differentiate recurrent and DP genotypes. Coverage can be assigned incrementally to each marker. Effective marker coverage for RP recovery is one marker every 15–20 cM. Thus, in a maize population with a total map length of 1,800 cM, approximately 100 equally distributed markers are appropriate. If marker-assisted selection is initiated in generations later than BC1, increased marker density should be considered since additional opportunity for recombination in between markers has occurred without selection pressure.

Expected gain in %RP can be made estimated through mathematical formula and simulation software. With mathematical formulae, population size can be derived from the variance estimates of BC populations, probabilities of picking the needed individuals from an expected selection tail, and the probabilities for identifying more than one plant in a selection tail. Alternatively simulation software may provide a more accurate answer of RP expectations with marker-assisted backcrossing. Parameters including population size, selection intensity, number of individuals advanced, and map positions of markers utilized can be use to predict gain over generations. Software such as these are more effective than mathematical formulas in predicting the actual outcome since the exact marker set and transgene position can be incorporated into the simulated population of recombined individuals.

Acknowledgments

I would like to acknowledge Dr. Mike Kerns and Dr. Wayne Kennard for their technical and editorial inputs during the writing of this paper.

References

1. Harlan, H. V., M. N. Pope. 1922. The use and value of back-crosses in small-grain breeding. *Journal of Heredity*, 13, 319–322.
2. Fehr, W. R. 1991. *Principles of Cultivar Development.* London: Macmillian.
3. Allard, R. W. 1960. *Principles of Plant Breeding.* New York, NY: Wiley.
4. Hospital, F. (2002). Marker-assisted back-cross breeding: a case-study in genotype-building theory. In Kang, M. S. (Ed.), *Quantitative Genetics, Genomics and Plant Breeding* (pp. 135–141). Oxon, UK: CAB International.
5. Babu, R., Nair, S. K., Prasanna, B. M., Gupta, H. S. (2004). Integrating marker-assisted selection in crop breeding – prospects and challenges. *Current Science*, 87(5), 607–619.
6. Visscher, P. M., Haley, C. S., Thompson, R. (1996). Marker-assisted introgression in backcross breeding programs. *Genetics*, 144, 1923–1932.

Appendix A

Application No. 08-107-102n

U.S. DEPARTMENT OF AGRICULTURE
ANIMAL AND PLANT HEALTH INSPECTION SERVICE
BIOTECHNOLOGY REGULATORY SERVICES

BRS NOTIFICATION - INTRODUCTION OF GENETICALLY ENGINEERED PLANTS

1. NAME, ADDRESS, TELEPHONE, AND EMAIL OF APPLICANT

Name: Dr. Marvin P Scott
Position:
Organization: United States Department of Agriculture/Agricultural Research Service
Organization Unique ID:
Address: 1407 Agronomy Hall
Ames, IA 50011
County/Province:
Township/Island:

Day Telephone: 515-294-7825
FAX:
Alternate:

Email 1: paul.scott@ars.usda.gov
Email 2:

2. INTRODUCTION TYPE

- [ ] Importation
- [ ] Interstate Movement
- [ ] Interstate Movement and Release
- [x] Release

3. APPLICANT REFERENCE NUMBER

Scott2008

4. CONFIDENTIAL BUSINESS INFORMATION VERIFICATION (CBI)

Does this application contain CBI? [ ] Yes [x] No

CBI Justification:
N/A

5. REGULATED ARTICLE

Scientific Name: Zea mays
Common Name: Corn
Cultivar and/or Breeding Line: B73, Mo17, B110, Va26, BS11, BS31

6. PHENOTYPIC DESIGNATION

1) Phenotypic Designation Name: 19 zn GFP
Identifying Line(s):
Construct(s): P228
Mode of Transformation: Biolistic

Phenotype(s)

MG - Visual marker

Genotype(s)

Gene(s) of Interest

Promoter: Transcriptional control **from** Zea mays - 19 kDa zein promoter

coding sequence: encodes protein **from** Zea Mays - GFP coding seuqence from Aequorea victoria

Terminator: Transcriptional termination **from** Zea Mays - Nos terminator from Agrobacterium tumefaciens

Selectable Marker

Gene: Bar gene encoding bialaphos resistance **from** Streptomyces hygroscopicus - Maize ubi-1 promoter - S. hygroscopicus Bar CDS - Nos Terminator

2) Phenotypic Designation Name: 27 zn GFP
Identifying Line(s):
Construct(s): P230
Mode of Transformation: Biolistic

Phenotype(s)

MG - Visual marker

Genotype(s)

Gene(s) of Interest

Gene: Production of GFP **from** Zea mays, Aqueora victoria, Agrobacterium tumefaciaciens - Maize 27 kDa zein promoter - A. victoria GFP CDS - A. tumefaciens nos terminator

Selectable marker

Gene: herbicide resistnace **from** Zea mays; Streptomyces hygroscopicus; Agrobacterium tumefaciens - Maize ubi-1 Promoter - S. Hygroscopicus BAR CDS - A. tumefaciens nos terminator

3) Phenotypic Designation Name: glb1 GFP
Identifying Line(s):
Construct(s): P231
Mode of Transformation: Biolistic

Phenotype(s)

MG - Visual marker

Genotype(s)

Gene of interest

Gene: Production of GFP in embryo **from** Zea mays; Aqueora victoria; Agrobacterium tumefaciens - Maize Glb1 promter - A. victoria GFO CDS - A. tumefaciens nos terminator

Selectable Marker

Gene: Herbicide resistance **from** Zea mays; Streptomyces hygroscopicus; Agrobacterium tumefaciens - Maize ubi-1 Promoter - S. Hygroscopicus BAR CDS - A. tumefaciens nos terminator

4) Phenotypic Designation Name: zmhb-GFP
Identifying Line(s):
Construct(s): P284
Mode of Transformation: Biolistic

Phenotype(s)

PQ - Iron absorption enhanced
MG - Visual marker

Genotype(s)

Gene of interest

Gene: Production of hemoglobin fused to GFP **from** Zea mays; Aqueora victoria; Agrobacterium tumefaciens - Maize 27 kDa zein promter - Miaze hb CDS - A. victoria GFO CDS - A. tumefaciens nos terminator

Selectable marker

Gene: herbicide reisistance **from** Zea mays; Streptomyces hygroscopicus; Agrobacterium tumefaciens - Maize ubi-1 Promoter - S. Hygroscopicus BAR CDS - A. tumefaciens nos terminator

5) Phenotypic Designation Name: ZmHb
Identifying Line(s):
Construct(s): P331
Mode of Transformation: Biolistic

Phenotype(s)

PQ - Iron absorption enhanced

Genotype(s)

Gene of interest

Gene: production of maize hemoglobin **from** Zea mays; Agrobacterium tumefaciens - Maize 27 kDa zein promter - Miaze hb CDS - A. tumefaciens nos terminator

Marker gene

Gene: Herbicide resistance **from** Zea mays; Streptomyces hygroscopicus; Agrobacterium tumefaciens - Maize ubi-1 Promoter - S. Hygroscopicus BAR CDS - A. tumefaciens nos terminator

6) Phenotypic Designation Name: Lactalbumin
Identifying Line(s):
Construct(s): P64
Mode of Transformation: Biolistic

Phenotype(s)

PQ - Lysine level increased

Genotype(s)

Gene of interenst

Gene: Produces porcine alpha lactalbumin **from** Zea mays; Agrobacterium tumefaciens - Maize 27 kDa zein promter - synthetic alpha lactalbumin CDS - - A. tumefaciens nos terminator

Selectable marker

Gene: herbicide resistance **from** Zea mays; Streptomyces hygroscopicus; Agrobacterium tumefaciens - Maize ubi-1 Promoter - S. Hygroscopicus BAR CDS - A. tumefaciens nos terminator

Application No. 08-107-102n

7. INTRODUCTION

Release Site

Location Name & Description	Location Address	Contact(s)
1) Iowa State University Transgenic Nursery - Woodruff Farm	R38 and Zumwalt Station road Iowa **County:** Story **Proposed Field Test Start Date:** 5/16/2008 **Proposed Field Test End Date:** 5/15/2009 **Number of Proposed Plantings:** 3 **Quantity:** 500 Plants **Planting Comments:** The plot is 144,400 square feet or .003 Acres.	**1)** Paul Scott **Day Telephone:** 515 294 7825 **Email 1:** paul.scott@ars.usda.gov

Location GPS Coordinates: NW: -93.69185638771556,41.98406482890459,0 NE: -93.69043319558554,41.98403179521117,0 SW: -93.69183599354643,41.98296359724991,0 SE: -93.69043657426732,41.98298319306124,0

Release Site History: This site has been in cropping systems for about 80 years.

Critical Habitat Involved?: ___ Yes _X_ No

8. ADDITIONAL INFORMATION

Application No. 08-107-102n

I, *Marvin P Scott*, certify that the regulated article will be introduced in accordance with the eligibility criteria and the performance standards set forth in 7 CFR 340.3. The above information is true to the best of our knowledge.

I acknowledge this is not an application to move or import select agents, the genes expressing select agents, or the toxins made by the select agents, as described in 9 CFR 121.

If there are any changes to the information disclosed in this application, I will contact APHIS.

9. SIGNATURE OF RESPONSIBLE PERSON	10. DATE
Marvin P Scott	April 16, 2008

Appendix B

Performance Standards for Handling Transgenic Corn Released to the Environment

Scott laboratory group, 2008

1. Shipping and Maintenance at Destination

Transgenic seeds shall be placed in seed packets (manila coin envelopes) labeled with the stock source, and/or field and row locator number. Packets shall be placed inside a zip-lock or similar type of plastic bag. The plastic bags containing seed packets shall be placed inside two nested metal containers with packing material between them and these containers will be placed in a sturdy plastic or corrugated cardboard box with lid. Boxes shall be labeled with the USDA-APHIS notification number.

Certification:
Shipping container conforms to guidelines.

____

Signature Date

2. Inadvertent Mixing of Materials in Environmental Releases

Transgenic corn plants will be grown in isolation from nontransgenic corn in a field that is at least 660 ft. (1/8 of a mile) in any direction from other fields of corn. A border of at least four rows of sweet sorghum will be planted around the transgenic plot as a physical barrier to reduce pollen dispersal outside of the transgenic plot. An unplanted alley will separate the defined planting area from other plants. The width of this alley will be wide enough to allow for machine movement without mechanical mixing. This area will be periodically mowed to control weeds. Field plot maps, seed packet labels, plant labels, and harvest labels will be generated using our Prism database to minimize the chances of errors in planting and harvesting. Seed packets of transgenic material will be stored separately from packets containing nontransgenic seeds in labeled storage containers. Planters will be inspected following planting to ensure no seeds remain in them after planting.

Certification:
Planters inspected:

____

Signature   Date

3. Identity and Devitalization

Each plant will be labeled with a tag indicating the pedigree of the plant. Tassels that are not required for experiments will be removed prior to pollen-shed or covered during pollen-shed to minimize the chances of accidental pollination by transgenic pollen. Field will be monitored daily during the normal silking period for corn to verify that tassels are covered and bags are intact.

Vegetative material from transgenic plots will be devitalized by incorporation into the soil by chopping, plowing, disking, and/or other accepted cultural practice, and exposure to the elements. In the following year, the area will be planted to a crop other than corn (e.g., soybeans). The area will be monitored for the presence of volunteer corn plants, and such volunteers will be destroyed.

4. Viable Vector Agents

There are no viable vector agents associated with our experiments.

5. Persistence in the Environment

All seeds from transgenic plants and plants crossed with transgenic plants will be collected by harvesting whole ears in paper bags and placing these bags in mesh bags. Following harvest, the plot will be tilled to ensure rapid decomposition of the crop residue.

6. Volunteer Plants

Plants will only be planted in test sites that are accessible and will not be planted to corn in the following year. Each test site used in the previous year will be scouted for volunteer plants in May, June, July, August, September, and October. If volunteer plants are identified, they will be removed before flowering so that the volunteer plants do not reach maturity.

Certification:
Volunteer corn removed from 2007 site and within 660 ft. of 2008 site in May

____

Signature Date

Certification:
Volunteer corn removed from 2007 site and within 660 ft. of 2008 site in June

____

Signature Date

Certification:
Volunteer corn removed from 2007 site and within 660 ft. of 2008 site in July

____

Signature Date

Certification:
Volunteer corn removed from 2007 site and within 660 ft. of 2008 site in August

____

Signature Date

Certification:
Volunteer corn removed from 2007 site and within 660 ft. of 2008 site in September

____

Signature Date

Certification:
Volunteer corn removed from 2007 site and within 660 ft. of 2008 site in October

____

Signature Date

Index